Superhuman

Robert Winston and
Lori Oliwenstein

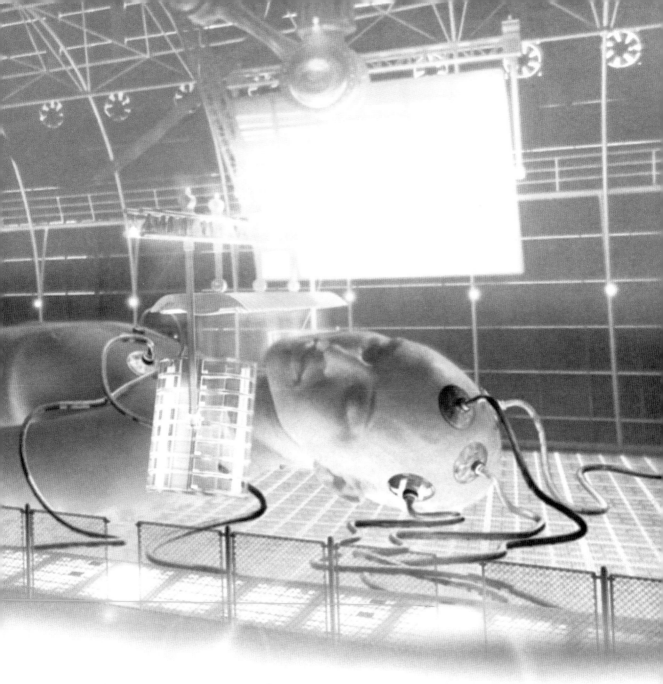

Superhuman

London, New York, Sydney, Dehli, Paris,
Munich, and Johannesburg

Publisher Sean Moore
Editorial director LaVonne Carlson
Art director Tina Vaughan
Project editor Barbara Minton
Editors Connie Robinson, Jill Hamilton
Art editors Gus Yoo, Michelle Baxter
Production editor David Proffit

ISBN: 0-7894-6306-7

First US edition published in 2000 by
Dorling Kindersley Publishing, Inc.
95 Madison Avenue
New York, New York 10016

This book is published to accompany the television series *Superhuman* which
was produced by the BBC and shown in the US by The Learning Channel.

Executive producer: Michael Moseley Series producer: David Hickman
Producers: Natasha Bondy, Judith Bunting, John Groom,
Liesel Evans, Kate Barker, and Dinah Lord.

Published by First published 2000 by
BBC Worldwide Limited,
Woodlands,
80 Wood Lane, London W12 0TT

Editorial Director: Shirley Patton
Project Editor: Helena Caldon   Art Direction: Linda Blakemore
Book Design: Rachel Hardman Carter   Picture Research: Miriam Hyman

Set in Sabon and Gill Sans
Printed and bound in Great Britain by Butler & Tanner Limited, Frome
Color separation by Radstoch Reproduction Limited, Frome
Jacket printed by Lawrence Allen Limited, Weston-super-Mare

See our complete catalog at
www.dk.com

Contents

Preface

The book accompanies the BBC series, *Superhuman*. The BBC science department is probably unique in television for the high quality of its output and the humanity of its purpose. The authors are proud to be associated with it, and with the Learning Channel, which has been such a fine coproducer. The book is written in the first person because it is essentially seen through my eyes as the presenter of the TV series. Nevertheless, it has two authors, and it has been a pleasure for me to work with Lori Oliwenstein on this project.

This book is a conflation—the result of huge effort by a large number of people. We are in debt to those brave men and women whose medical stories we tell, and we have the deepest respect for their courage and openness. It is also good to pay tribute to the many experts, physicians and scientists, whose pioneering work this book describes and who gave us their time so freely.

It has been a pleasure to work with such a talented production team. Television production is one of the highest forms of art in the modern world: It combines the undertaking of high-quality research, the ability to find striking and beautiful images, skill at putting vast amounts of material in order and selecting what is relevant, the blending of music and language to tell a complex story, a sense of theater, and the establishing of a rhythm that gives excitement on screen. It is an ephemeral art, but television images can, on rare occasions, make deep impressions and change people's perception of the world. We hope that this book makes the work of the team who made these programs a little less ephemeral.

Numerous people deserve special mention. Firstly, the producers who work under incredible pressure, and who in this series were sometimes brought in late to make these programs—increasing the pressures even more: Kate Barker made the program on cancer—her work is of the highest quality; Liesel Evans, the program on infection—a delight to work with an old friend from *The Human Body*; Judith Bunting, the program on transplantation—her delightful touch was also present in *The Secret Life of Twins*, the series with which I had the privilege to help; John Groom, the program on self-repair—John another old friend from *The Human Body* with whom it is always a real pleasure to work; Natasha Bondy, the program on trauma—her vibrancy shines through in her films; Dinah Lord, the program on fertility—her reputation as a film-maker is totally deserved. It was a privilege to walk in their footsteps.

Three talented researchers/assistant producers who helped so much with parts of this book and who were such a support during filming must be singled out: Rachel Hellings, Michael Lachmann, and Alison Dillon. Their speed of work, enthusiasm, intelligence, and commitment was extraordinary, and they were always helpful and good humored.

Michael Mosley, the executive producer, gave unstinting help in his charming way and was responsible for many of the big ideas in the book. David Hickman, the series producer, is an extraordinary filmmaker with a riveting pictorial sense. We are deeply grateful to him both for his artistry and for his help in assembling much of the material for this book.

Many of the pictures in this book were taken by the superb camera-men who worked on the series; they include Michael Coles, Ian Selvage, Neil Harvey, and my old friend Chris Hartley. It was also a pleasure to work with numerous skilled sound recordists, but especially George Hitchens and Keith Silva who were so much part of the production. David Barlow is a unique photographer with his endoscopes, photographic gadgets, and special lenses. His distinguished international reputation is thoroughly deserved. The computer graphics, which make some of the telling images in the book, were the remarkable creation of Rick Leary and his extraordinarily talented team.

I would like to thank my close friend and tireless agent Maggie Pearlstine and her assistant John Oates. They both kept a constant eye on the progress of this book and continually came up with sensible advice and reassurance. I am also most grateful to those at BBC Worldwide Publishing—Shirley Patton, Helena Caldon, Linda Blakemore, Miriam Hyman, and others—who have done so much to ensure the quality of this book, and who showed such forbearance during its production. I am also grateful to my son, Joel Winston, who read and made comments on parts of the manuscript between final exams at Cambridge.

Finally, without the remarkable Leo Singer this book would have been impossible. He contributed hugely to the text and to much new research, as well as checking older material. His clarity and speed at honing, refining, revising, and reordering whole areas have improved the text immeasurably.

Lori Oliwenstein would like to acknowledge a huge debt of gratitude to her agent, Leslie Daniels, of the Joy Harris Literary Agency, whose advice was always just what was needed, and whose uncanny knack of always knowing just the right thing to say has been invaluable. She owes another debt to her brother-in-law, Jeffrey Kluger, who not only led her to Leslie, but also let her bounce ideas and anxieties off him on a regular basis. She thanks Brenda Maceo and the staff at the University of Southern California's Health Sciences Public Relations office, who gave her wide berth as she sat hunched over her keyboard on a multitude of "book days," but always made sure to include her in their lunch plans.

But, most of all, Lori wants to thank her husband, Garry Kluger, a man whose stores of love and patience are truly superhuman, and her daughter, Emily Kluger, who is a daily reminder of what the superhuman potential is all about.

There are two other people I wish to acknowledge. Firstly, my friend Richard Dale—an ace filmmaker, always there at the end of a portable telephone for advice on all matters. Secondly, my wife Lira, whose tolerance, forbearance, support, and love were always in evidence even when I was disgracefully distracted during this project.

Robert Winston
London, 2000

Introduction

September 1999. We are filming in Greece—the island of Kos, just two miles from Turkish waters. It's the place, of course, made famous by Hippocrates who founded his medical temple here. It's a barren, stony landscape, in places a cross between the Mojave and the valleys around Jerusalem. It is dotted with shrubs and olive trees and is lit by a curious, limpid light. That light at 5 P.M., when we arrived, had a touch of indigo. Shortly after, my bedroom furniture trembled and rattled. It seems completely unreal that a short distance away, a massive earthquake has just occurred in Athens, bringing death by trauma and the risk of infection in the water supply. We have simply been sitting by the swimming pool of this hotel, drinking beer and watching the horizon turn more and more purple as the sun sets.

The people came to Kos by the thousands, medical pilgrims seeking a cure for their ailments. What they found was a healing temple, one of several dedicated to the god Asclepias, son of a mortal mother and the god Apollo. Within its stone walls, held aloft on the gently rolling hills of this Greek island and caressed by its mineral springs, the hopeful partook in a ritual called incubation or temple sleep. Their belief was that healing would come to them in a dream, a divine intervention brought forth by Asclepias himself. Or so they claimed.

In the days before Salk or Pasteur, before computerized body scans and electrical heart tracing, before antibiotics and gene therapy, there was

OPPOSITE *The ruins of the healing temple dedicated to the god Asclepias in Asklepion, Kos. Hippocrates was said to have practiced medicine here.*

ABOVE *A portrait bust of Hippocrates, founder of our modern system of medicine.*
BELOW *The 13th century method for relieving pressure on the brain by drilling a hole in the skull.*

little more a healer could do than give a body the chance to save itself. In ancient Greece and Rome, where gods eclipsed scientists, and the study of humans was considered a base pursuit, this was the best that medicine had to offer. Our modern system of medicine dates back to Hippocrates, who was born on the island of Kos and there established his medical school. And yet Hippocrates himself is said not only to have been a descendant of Asclepias, but to have trained in that god's healing temple.

Hippocrates made little use of drugs. He did not try to interfere with nature. He knew that most diseases had a natural tendency to cure themselves: "Our natures," he said, "are physicians of our diseases." Hippocrates used fomentations and bathing, diet and advice. Occasionally, he insisted on fasting, but fasting was always carefully controlled and was never done to excess. Mostly the diet was very simple and included vinegar and honey. Grape vines and beekeeping are still much in evidence on Kos today.

Hippocrates also engaged in basic surgery, using trephining to drill holes in the skull for serious head injuries, tar on wounds that might suppurate (it is now known to have antiseptic properties), and splints for setting broken bones. He stressed the importance of cleanliness during surgery.

Today, of course, we have much more technology. We have machines, assays, and pharmaceuticals. More importantly, we have built a system of medicine that relies on scientific inquiry. We try to find both the basis of disease and the best way to abolish it. We no longer need to dream our cures; we have the means to make them happen.

Still, everything we have in the present is built to some extent on the past. Good modern physicians—Hippocratic physicians—recognize the debt that today owes to yesterday. They are willing to go back to our philosophical roots and rediscover the body's innate wisdom, and alloy that with the medical wisdom we have accumulated over the centuries. Which is why, of course, this book frequently harks back to the historical perspective.

The modern physician realizes that the only way to heal the body is to enlist its help and take its internal counsel. Harnessing the body's capacity to heal itself is an idea so ancient that it's modern, a concept so simple it is almost revolutionary.

We often forget to listen to our bodies when we are well. Only when we fall ill do we become physiologically self-absorbed, acutely aware of every ache and twinge. But the healthy body has much to tell us. We need to understand precisely how the human body works, in sickness and in health, in order to reach our full healing potential.

The human body knows best how to keep itself in equilibrium, and how to right itself when that equilibrium is disrupted. We have only just begun to understand and use those restorative secrets in tandem with our own medical techniques. Medicine is at its best when it works *with* the body, expanding and exploiting its myriad resources by using intelligent methods and technology. Learning how to take advantage of our finely tuned physiology, the province of Superhuman medicine, can help us be less mortal, less prone to the flaws evolution has left behind. It can make us stronger, healthier, better. It can bring out the Superhuman.

I must emphasize that none of the above is a plea for so-called alternative medicine. I have great difficulty, as do most responsible physicians, with medical practice that is based on magic, unsupported belief, pure ritual, or hair-brained theory. I have nothing against challenges to conventional medicine, but medical practice must be based on evidence.

Of course, there are many alternative medical practices—acupuncture and herbalism, for example—that can now be shown to be based on a good physiological rationale. Acupuncture works by releasing opiate substances in the brain; and many herbal remedies have a pharmacological basis. For instance foxglove can be shown to have a direct action on the conducting system of the heart. But it is unethical to give a patient a potion that you think might work, but which has not been tested. The wrong treatment may not only *not* heal, but may cause side effects. Even if it is harmless in itself, it may prevent more effective treatment from being promptly administered.

Of course, that is not to say that many treatments have been found to produce an effect empirically. But reliance on empiricism without inquiry is no longer acceptable.

It is worth considering why alternatives to conventional medicine are so popular. It is an interesting question, but surely proof of their efficacy is not one of them. Perhaps it is because practitioners of alternative medicine give their patients more time, just as Hippocrates did on the island of Kos. Perhaps it is because alternative medical remedies often have a powerful placebo effect; if you believe strongly that something will do you good, it *will* make you feel good if you take it. If you feel good, you are more likely to get better. Perhaps, after all, alternative medical practice has a flavor of the Superhuman, the recognition of the human body's ability to heal itself.

There is a common view among so-called futurists, self-appointed prophets who are prone to attaching labels to periods of time, that the 20th century was the century of physics, and the 21st century will be the century of biotechnology.

This is, admittedly, one of their more reasonable notions. Think about the defining technologies and scientific highlights of the last century. In 1905, Einstein's Special Theory of Relativity quietly inflicted a deadly blow to the mechanical, commonsensical Newtonian universe. Ten years later, along came General Relativity and the inference that space is curved, a revolution that transformed our understanding of gravity and the interaction of celestial bodies and gave birth to the Big Bang and the expanding universe. Quantum theory threw a monkey wrench into the works of our understanding of the smallest and most fundamental particles of matter, as well as introducing worrisome ideas like the uncertainty principle.

More practically, these scientific—and philosophical—revolutions paved the way for the atomic bomb, the microchip, television, and the communications revolution. Very few aspects of our existence have been left untouched by these technologies.

In one respect, the biotech age started in earnest in 1953. At that time, as the world was going nuclear and the superpowers were digging in for the Cold War, biology was often seen as a barren science, condemned to dry description of anatomy and botany. Enter the DNA hunters: Francis Crick and James Watson (with the help of Rosalind Franklin, who unfortunately missed out on some of the glory). They finally cracked the structure of DNA—the double helix.

The structure of DNA, the key to life itself, was the Special Relativity of molecular biology. Move over, Schrödinger's cat and nuclear fusion. DNA is about what makes us human. The endless twine of the DNA double helix is an enormously complex biological code, made up of some 60,000 to 100,000 genes encoded by three billion chemical pairs. DNA allows our cells, and, therefore, the human being as a species, to divide and reproduce. DNA governs and regulates our growth from a single-celled fertilized egg to a fully developed human. DNA keeps us alive—by regenerating our tissue, producing thousands upon thousands of different hormones and proteins necessary for our body to thrive, and controlling the immune system to fight off invaders. DNA is the software on which we run, and the biological legacy we pass on to the next generation. If ever a discovery deserved a Nobel Prize, this was surely it.

The DNA revolution slowly permeated practically every branch of biology and every arm of the medical profession. Sixteen years after Crick and Watson constructed their elegant molecular model in a Cambridge laboratory, a Harvard Medical School team identified and isolated the first single gene, a tiny chunk of bacterial DNA involved in the metabolism of sugar.

In 1980, Martin Cline and his team created a transgenic mouse, which carried functional genes from an entirely different species. In 1982, the first genetically engineered drug, a form of human insulin produced by bacteria, was approved by the US Food and Drug Administration. The year 1990 saw the launch of the international Human Genome Project, and an American geneticist, W. French Anderson, introduced foreign genes into the DNA of a four-year-old girl with an immune-system disorder.

The stage was set for the biotech century. Biotechnology has become the dominant scientific powerhouse across the world. However we choose to make the comparison—counting newspaper headlines or adding up government grants, funding for private R&D, the numbers of researchers, shiny new paint jobs, and state-of-the-art new facilities— all these add up to an entire new industry.

Every week—or is it every day?—we hear seemingly impossible news from the cutting edge of molecular biology: Strokes in rats are treated with laboratory-grown brain cells; transplants of embryonic nerve cells have restored the rat's movement and behavioral function; a new generation of AIDS drugs, known as protease inhibitors, appearing to control the spread of HIV in the bloodstream to such an extent that contracting the virus may no longer be a certain death sentence; doctors launching a trial for a new male contraceptive pill;

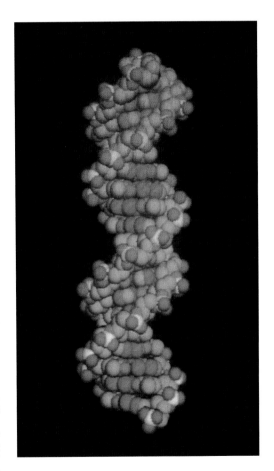

DNA is an extraordinary molecule which is tightly twisted, far more so than shown here in this computer generated graphic. The letters of the genetic code are in the center of the spiral.

scientists cloning veins for bypass operations; and the first all-human replacement part, a two-inch length of blood vessel, is implanted in a dog.

Science fiction writers have dreamed for years about the physical future of mankind. The idea of building something superhuman is even older; Mary Shelley's story of Frankenstein, who built a monster from old bones, was published in 1818. Much earlier, in the 5th century, Jewish writers posed the idea of a Golem, a giant made of clay who is given life by means of an amulet on which is written the holy name. More modern science fiction has dreamed of how we might combine physiology and technology to create a futuristic hybrid, a new breed of human that could perform feats of strength or physical prowess. And it is extraordinary that the so-called Terminator films, featuring a tougher, quicker, longer-lasting, and more physically and mentally agile bionic "human," are so popular.

One such book, Bruce Sterling's *Schismatrix*, supposes that a future mankind has split into two groups embracing two very different ways of life, ideologies, and technologies. Neither of these factions have been content with our standard range of human abilities and talents, and have stretched the limits of what it means to be human. Mankind has become two very different animals.

One of the groups, the "Mechs," employs artificial intelligence, bionics, and other technology to become more than human. Mechs represent a future that combines man with machine; this character has become a staple of Hollywood sci-fi. They have become bionic, replacing inefficient, weak, or unreliable body parts with metal and plastic. Artificial intelligence has enhanced or replaced humankind's feeble and capricious mental processes with the solidity and speed of silicon and binary code.

The other group, called the "Shapers," use biotechnology, genetic engineering, and subtle psychology to become superhuman. Shapers do not try to build a new human; they enhance and expand our already substantial array of talents. Instead of replacing a diseased part with an ersatz copy made of alloy of titanium or some Martian ore, as their archenemies the Mechs would attempt, they try and fix their own homemade biological organs. They use technology to transform humans, rather than turn them into technological beings.

OPPOSITE Dr. Peter Kyberd testing a myoelectric prosthetic hand he developed. Sensors detect electrical activity in the muscles of the wearer's forearm and a built-in computer tells the artificial hand to open, close or grip.

We once had dreams of a Mech-like existence. Throughout the 20th century we exercised an increasing talent for building precision mechanical

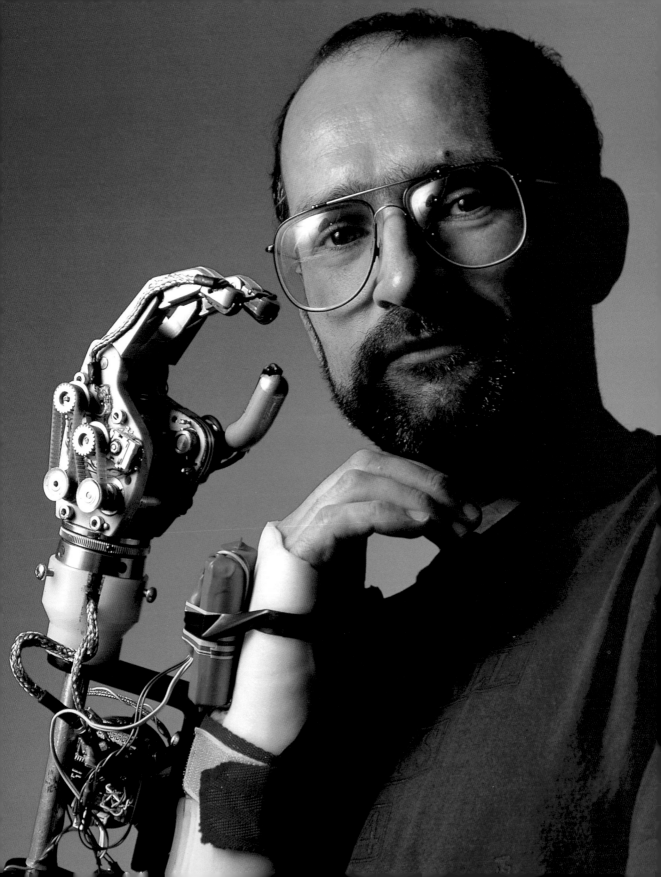

devices, machines that got smaller and smaller, devices that eventually could fit on a tiny piece of silicon just a few millimeters square. Lots of people thought the marriage of man with machine, the brawn of the Bionic Man with the brains of HAL, was the next step in humankind's evolution. But we have been disappointed.

We have tried, and will keep trying, to engineer artificial replacement parts, whether they are mechanical prosthetics to replace diseased limbs, advanced plastics for skin transplants or neurological add-ons for a diseased brain. But apart from certain relatively simple devices—for example, titanium hip joints or cochlear implants—we have overreached our capabilities. It has become apparent that our future is veering towards the "Shaper" scheme of things. Biotechnology, genetic engineering, and self-repair—these are the technologies that increasingly seem to be advancing the frontiers of medicine.

The idea of the Superhuman is a metaphor for what humans may become, not just a futuristic fantasy. We are seeing real, successful treatments and medical methods. Superhuman medicine uses human genes to fix human ills; it revs up the immune system so that the body itself may conquer a lethal virus. And if that is not enough, the Superhuman metaphor goes several steps further. The Superhuman being is also able to leap over the physiological hurdles set for us by evolution. She can have babies after menopause; he can regrow damaged nerve cells.

Being healthy and staying that way would be much easier if the human body did not have what appears to be a number of major physiological pitfalls. We are a well put together biological machine, but we are not perfect. There have been numerous compromises and concessions hammered out over millions of years of natural selection. In some respects we are creaky, patchy creations, all checks and balances and make do and mend. After all, evolution is about the survival of the fittest, not the survival of the perfect solution. Our genes encode for success, but carry messages for failure.

In many respects, we have outgrown our bodies. Some of our limitations are not simply the result of the *ad hoc* aimlessness of evolution; the structures we are living in were built to hunt and gather rather than to shop and microwave. They were built to stand on two legs, not to speed around on four wheels. And they were built to last only long enough to reach reproductive age—30,

> ... the structures we are living in were built to hunt and gather rather than to shop and microwave.

40 at the most—not to live 60, 80, 100 or more years—decades after we have passed on our genes to the next generation.

In the Stone Age, our biggest problems were those of immediate survival. Our bodies evolved to hoard fat and calories to deal with frequent bouts of famine. Now, it simply does not know what to do with today's high-fat, high-calorie, high-volume meals. Our nervous system could cope when we had to flee an approaching saber-toothed tiger or other prowling predator, but it is at a loss when forced to deal with the every-day, low-level stress of an industrialized world. And our reproductive systems were used to producing one, occasionally two, and on very rare occasions three babies at a time, but not used to being coaxed by drugs and surgical interventions to nurture seven or eight in a single womb.

Evolution has not had time to adjust our anatomy and tweak our physique so that we may better handle the pressures of the modern world. That is not to say that 10,000 years ago, 100,000 years ago, we all lived happy, healthy, carefree lives. Stone Age man was susceptible to infection. He was killed by heart failure, congenital disorders, and immunodeficiency conditions, just like us. He rarely lived long enough to get cancer, but it was a threat all the same.

Despite Stone Age man's short life expectancy, the human body is better adapted for the nomadic life of this hunter-gatherer than it is for a modern city dweller. To reach the next level in healing this Stone Age organism, then, we have to work out ways to reconcile our evolutionary past with our technological present. We should understand and respect our body's limitations. Recently, for example, surgeons have begun to recognize that a body pulled from a car wreck has often reached the limits of its endurance. Doing anything more than is minimally necessary to keep the patient alive may actually result in losing the patient.

This does not mean, of course, that science and scientists should not stretch the limits that evolution has set down. They do this all the time. In cages in the world, supposedly paralyzed mice are scurrying about because researchers refuse to be constrained by conventional wisdom. But when these researchers push the frontiers of medical treatment, they need to push gently. The human body does not respond well to the brute force of modern technology. We need to make our medicine intelligent.

The researchers who populate the chapters that follow are attempting to do just that. They are using the talents embedded in our ancient physiology to perform feats that blunter, more traditional methods could never achieve.

There are six chapters in this book. Each depicts a major area of modern medicine and in each of these fields, I believe, the Superhuman has the greatest influence and potential. In Chapter One, we meet trauma surgeons who are learning that sometimes the best way to treat a trauma patient is to step backward. Emergency paramedics are discovering that it may not be best to overuse technology at an accident site, but rather to just pick up the patient and run. Physicians are finding that shock, a natural defense mechanism after a serious trauma and blood loss, can be helpful in saving the body and the brain.

Chapter Two demonstrates that a true Superhuman is built out of skin and bone and muscle and neurons, not titanium and plastic. The smart money is not in backing artificial parts. Flesh-and-blood organs and biological tissue beat any man-made device hands down, and the race is on to try and find new sources for biological organs and new ways to cope with the complex phenomenon of rejection. Some researchers are attempting to persuade the body's immune system to accept a donated organ as its own by finding ways of cheating the immune system. Other researchers, responding to the organ shortage, are hoping to find the solution in the animal kingdom.

Self-repair, or regeneration, is a holy grail of advanced medicine. Chapter Three introduces the scientists who are trying to reclaim our lost powers of regeneration—to help us build new blood vessels, regrow organs, and reseed the brain. They are attempting to engineer human bits and pieces using materials found in the body rather than fabricated in a factory. They are even exploring the idea of cloning human cells to create organs and tissues made to order for the people who most need them.

Chapter Four tells the story of a fierce war over one of humankind's most treacherous enemies—cancer. This is a battle that has raged for many years, and despite the many promises of the "cure around the corner," we are still taking potshots from the trenches. But we are starting to understand how cancer turns our own cells against us, inciting a cellular insurrection. The researchers in this chapter are valiantly trying to discover the best ways to regain control over these cellular rebels.

In Chapter Five, we investigate another frontline battle. Experts in infectious diseases are searching for new, winning strategies in the arms race between man and microbe. They hope to find an Achilles heel to exploit in bugs that are already resistant to almost every man-made drug. They are also trying to take some of our greatest microbial enemies and turn them into allies. We discover why Kenyan prostitutes, who are immune to HIV, are the inspiration for research into a vaccine for AIDS.

Chapter Six explores the world of fertility and genetic manipulation. Medical innovation in these areas generates a large amount of debate and an even larger amount of fear. I hope to argue that the technology has a number of indisputably positive applications and that the dangers are exaggerated, whether these involve a couple avoiding the risk of conceiving a child with cystic fibrosis, or the freezing of immature eggs to expand the window of a woman's fertility. But this area of research could conceivably give us the power to alter the course of human evolution—to make a flesh-and-blood Golem, a Superhuman that is totally new. Therein lie the tremendous benefits and the potential dangers.

Medicine has not done all it set out to do. We have yet to conquer most forms of cancer. The AIDS virus continues to spread, and kill. In addition, the price that we pay for our accomplishments is landing us in a morass of ethical quandaries. Is it fair to use animals' organs to save human lives? Should we really be picking and choosing our embryos, our future children, based on the genes they do or do not carry? Can we justify testing potentially lifesaving drugs on unconscious trauma patients who are unable to give informed consent? Should we allow aborted fetuses to be used as the source of curative stem cells, or create human clones from which cells or organs could be harvested for therapeutic use?

Becoming superhuman has some literal costs as well. The implementation of the latest medical technologies and the basic research that precedes them are likely to consume huge chunks of our resources. Can our society afford to provide all that is promised by those advances?

The answer to that question is almost certainly no. Inevitably a large proportion of medical research is either conducted or funded by private companies. They are not charities. No one expects these entrepreneurs and venture capitalists to spend millions on an enterprise designed to enhance human happiness. That is not how private enterprise works, notwithstanding the lip service that some chief executives and biotech public relations people pay to the idea of public service.

Of course, one has to accept that without the promise of profit, however far in the future, the research that these companies conduct simply would not happen. That is the nature of the system, and biotech money is by no means easy money. These companies' mixed fortunes on the stock markets show the degree of risk which these ventures involve, and the nervousness which investors continually display.

But the commercialization of Superhuman medicine presents the question of access to treatment. It goes without saying that most research, trials, and treatment are limited to the developed world for the moment, and are more particularly concentrated in the United States. Many third world nations cannot even afford the medical technology of a previous generation—bypass surgery, antibiotics, and preventive screening, all of which we now take for granted in the developed world.

Biotech capitalism means that only a select few of us become healthier, at least in the short or medium term. The divide between the medical haves and have-nots is most likely to grow ever wider. Are we about to enter the era of the Superhumans versus the Stone Agers?

Superhuman medicine does not just cure diseases, combat infection, or save a crash victim. It also allows us to extend life expectancy. Medicine has been successfully increasing our time on this earth for decades. Now, with a more refined knowledge of how and why we age, and the molecular mechanisms that cause us to die, there are researchers who hope to slow down our biological clocks—even shut them down—or, in the extreme, reverse the aging process completely.

This process of extending the human life span is likely to continue, especially if we can replace or regenerate ailing body parts to extend their life even further. Tissue engineering, gene therapy for cancer, genetic manipulation of the germ cells—the eggs or sperm—all these technologies may give us the ability to live 100, 120, even 150 years. In the 22nd century we are quite likely to be part of a population that consists mainly of people older than 80.

Worried? Perhaps we should be. Even if we ignore the fact that half the country may have to be turned into golf courses to accommodate lively retirees, we may have to accept that birthrates will need to drop substantially to keep the population at a reasonable level. We may, in fact, even impose limits on the number of babies born.

In addition to this prospect, if the technology is publicly funded and thus widely available, the money will come from the working population. This minority group, young enough to work and pay taxes, may not take kindly to funding the continuing health of a graying population that shows little sign of making way for the next generation.

However, it is more likely that the only people who will get this shot at immortality are the ones who can afford to pay privately. Public funds will simply not stretch to keeping us alive for an extra 50 or 60 years. The rest of us may have to be content with little more than our three score years and ten (or perhaps twenty). This can only deepen the social and economic divide. Those who pay get to stay. The rest of us must be content with our mortality.

Modern medicine has a yen for speed. The pace of invention and discovery has increased substantially in the past 20 years. For us passengers it can be a terrifying ride. Medical advances cannot always pause for deep ethical or practical consideration. They introduce notions which, only 50 years ago, would have seemed not only bizarre, but positively inhuman. Our ethics frequently follow our capabilities.

But the headlines keep on coming. In fact it is easy to become complacent, desensitized, bored even. Worse still, we become immune to these startling discoveries and, at the same time, expectant of the success of the experiments. In the following chapters I hope to demonstrate the invention and technical prowess these achievements display. All of these inventions, trials, and therapies involved tremendous human effort by the team leaders and their unsung researchers who beaver away behind the scenes. The work takes years of painstaking research, experiment, and analysis, and encounters endless false leads and countless dead ends.

The scientists, however, are nothing without the human guinea pigs with whom they work. In this book, we meet dozens of them: the man who survives an industrial accident despite burns over much of his body; the woman who struggles to regain sensation in her paralyzed legs; the prostitute who manages to hold the human immunodeficiency virus at bay; the golf enthusiast who succumbs to the ravages of cancer only after the fiercest of battles.

Some scientists are successful, some are not. All of them are instrumental in extending the boundaries of medical knowledge. The *Superhuman* journeys along the cutting edge of medical innovation and places the technology in the context of past achievements, future possibilities, and the ethical dilemmas that rapidly arise from these ambitious projects.

Scientists may be trying to take us from the merely human to the Superhuman, but in the end it is our humanity that defines us. This was clearly the guiding principle of Hippocrates on this island of Kos where I sit writing tonight: "Our natures are the physicians of our diseases." If we can exploit our extraordinary capacities to cure ourselves without compromising what it means to be human, then we will become true Superhumans.

I Trauma

May 1996. It was the end of the Memorial Day weekend, and it felt like the beginning of summer. High school graduation was just two weeks away. Twenty-one-year-old Jennifer Vaughn and her friend Melissa were driving aimlessly through the small town of Everett, Washington, looking for friends who might be out enjoying the sunshine.

At the edge of the small town, on a quiet residential road, they decided to turn back and head toward home. Exactly what happened next is still in dispute, but it is clear that, as Jennifer swung her Honda Civic around in a U-turn, a VW Jetta, which police later estimated to be traveling at 70 miles per hour, slammed into the driver's side of the Honda. What was also clear was the imprint of the number plate in the metal of Jennifer's car door.

Melissa was thrown free of the car and bounced off a chain-link fence some 50 feet away. She was lucky. She broke a few bones and recovered quickly. Jennifer, on the other hand, was trapped inside the mangled crush of metal. Firemen had to saw off the roof of the car before paramedics could reach her, and, when they saw her, they knew she was in serious trouble. The accident left Jennifer with a fractured pelvis, bleeding liver, ruptured spleen, and massive internal bleeding. She was unconscious and in shock. Jennifer's life was on a knife edge; after she was airlifted to a hospital, her chances of survival were just fifty percent.

When she arrived at Harborview Medical Center, Jennifer had two teams of trauma specialists trying to save her life. The first team consisted of the hospital doctors, with their panoply of hi-tech sensors and regulators, artificial heart machines and intravenous lines, and cabinets full of fluids, blood products, and pharmaceuticals. The second consisted of her own body's defense mechanisms, the automatic and highly complex reactions that kick in when the body is in serious trouble.

But the two teams do not always cooperate. The body is not designed to cope with being mangled by a one-ton hunk of metal traveling at 70 miles per hour. So our genetically programmed response will try to save us, but our body can overreact and set off dangerous and sometimes fatal chain reactions. If too much blood has been lost, or the torn flesh and ruptured organs are too damaged to repair, then our ability to stem the wounds and regenerate damaged tissue are overwhelmed. Physicians, with the tools and techno-wizardry of modern medicine at their disposal, do their best to offset those reactions, keep the body under control, and restrict damage to a minimum.

However, over the past few years doctors have become aware that modern trauma techniques are sometimes directly *opposed* to our natural physiological response. In many cases it seems we are not taking advantage of the intelligence of our own natural defenses. Sometimes, in our ignorance, we may be seriously diminishing our patients' chances of survival with the crude and unthinking use of modern medical technology.

Over the past century, familiar procedures that make up modern trauma medicine have been developed. They are routinely applied by many doctors dealing with victims of serious trauma, and the paramedical teams called out in an initial response. But some experts are skeptical and now believe that some of these techniques, used daily in hospitals and in the field all over the world, are based on little more than habit.

This realization is starting to hit the doctors at the sharp end of trauma medicine, as well as the scientists and researchers who theorize and experiment in their laboratories. They are increasingly coming around to the idea that we should try much more to work with the body, not against it. They are exploring how our own defense mechanisms, refined over millions of years, and born out of confrontations with saber-toothed tigers and giant black bears, are teaching us a thing or two about modern trauma procedure.

Evolution has not left us with the automatic ability to survive any serious major trauma. If a saber-toothed tiger had ripped off my Stone Age leg, I would probably have bled to death. But if it only managed to drag its claws through a few layers of skin, my body might have been able to handle it.

Within seconds of the damage, hormones such as epinephrine (formally called adrenaline) would course through my veins, firing up nerves and muscles, preparing the body to make an escape. Blood flow to unessential organs, such as the stomach, would be diverted to my muscles, in an attempt to make sure that I could run if the cat turned on me for a second time. The pallor of my face through sheer fright would also be an effect of epinephrine, taking the blood supply away from my skin and conserving it for more emergency functions. My blood sugar would rise in response to its release from the tissues, so that there would be a readily usable source of energy for immediate consumption.

However, if the damage was really serious, I would not be running away. I would be dead. The biggest immediate problem would be severe blood loss, meaning that the body would be left without the ability for damage control. Without blood, tissues and cells have no oxygen, and consequently no chance of survival regardless of our fitness or strength. So one of the first emergency functions would be to stanch the leak.

My platelets, which make up around five percent of all the cells in my bloodstream, would help with the formation of an initial plug, and the protein, fibrin, would be rapidly deposited to weave a web of sticky fibers to form a clot. Within minutes, although rather less urgently, thousands of white blood cells would be rushed to the site to help clean up the mess. A large part of their job would be to target any foreign invaders, such as bacteria, that might have found their way through the breach in my defenses.

Initially, I might feel very little in the way of pain. Pain could be a distraction in such an emergency, when I might need all my strength for flight or possible fight. But as blood loss, pain, and shock set in, my blood pressure would automatically drop. This is, in part, the body's attempt to conserve as much as possible of the 10.5 pints (five liters) of blood that remain in circulation.

Evolution has certainly not equipped us with the means to survive everything that modern life can throw at us. It has not given us the ability to walk away from a brutal car crash, never having factored the automobile into our physiological fittings. It could not anticipate that we might take up skateboarding, climb sheer cliff faces, or manufacture corrosive acids. Nor could evolution fully prepare us for the internal damage that might follow the ricochet of a bullet inside the chest or abdomen.

Enter the modern trauma medic, with all the glamour and grit of the profession. Many people find it difficult not to admire the sheer physicality of the world of trauma medicine. Decisions must be made in a matter of seconds and the stakes are of the highest. For those patching up the wounded in hospital emergency rooms, there is nothing more physical than the blood-and-guts of trauma.

For those patching up the wounded in emergency departments, there is nothing more physical than the blood-and-guts of trauma … Trauma is undeniably epinephrine-rushing, pulse-pounding, in-your-face medicine.

Cancer, heart disease, and viral infection frequently work silently and secretly, commandeering or destroying cells, blocking up blood vessels. They can sometimes remain unseen and untreated for months, if not years. On the other hand, there is no hiding from a chemical burn, a lacerated lung, or a cracked skull. Trauma is undeniably epinephrine-rushing, pulse-pounding, in-your-face medicine.

According to television screenwriters, trauma doctors are the doers, not the thinkers, of the medical world. Their role model is a square-jawed action hero whose virtue lies in speed, guts, and instinct rather than a

considered, thoughtful intelligence. The heroic moment in medical melodrama is almost always a piece of spur-of-the-moment bravery; at the very least, those who play by the rules get in the way, and, at the very worst, they end up letting their patients die.

It is not only the patient who has a surge of epinephrine after trauma. The stress experienced by medical and paramedical personnel can easily lead to the wrong decision in a dire emergency.

Reality, of course, bears just a slight resemblance. Perhaps the doctors are not as good-looking, but the best ones go about their work with an astonishing degree of cool and calming professionalism, even in the most trying and stressful circumstances. This calm comes mainly from an ordered and well-established protocol of how to treat a trauma patient. Television screenwriters would have us believe otherwise, but the majority of trauma work is structured by rules and techniques developed over years of experience.

The word *trauma*, from the Greek word for wound, covers a multitude of minor and major problems, starting with the most innocuous—cutting one's finger while slicing onions, burning a hand on a hot iron, breaking a bone, and going all the way to stabbings, shootings, or being hit by a car. All trauma involves some kind of physical injury.

In fact, the treatment of trauma was likely to have been the very first form of medicine attempted by prehistoric man. Abdominal pain, infection, or chronic fatigue would have been a mystery; these were conditions which might be resolved by time, or some obscure and hard-to-find herb, or even by witchcraft or magic. On the other hand, traumatic injury is highly visible. It is easy to diagnose a missing limb, and profuse bleeding is something which, self-evidently, is not supposed to happen. Such conditions demand urgent attention. Unfortunately, we do not know how effective the Stone Age treatment might have been, but we can make an informed guess from studying those few very primitive societies living in remote parts of the world today.

Skull trephining in Peru using a flint as a scalpel must have been a rather uncomfortable experience for the unfortunate patient.

The first hint of trauma procedure in very ancient times comes from skulls dating back to the neolithic age. Numerous skulls that have now been recovered from a wide range of archaeological sites in Europe—from France, Austria, Poland, Russia, and Spain—all show a characteristic injury in common. All have neat holes drilled in the cranium. Almost certainly, these holes were caused by a flint stone which was used as a key surgical instrument. Trephining has long been a medical practice and seems to have been employed to let out "evil spirits" or after a person had suffered a major head injury, typically a skull fracture.

Head injury must have been comparatively common in prehistoric times, and it is interesting that some of these skulls show other injuries, which were probably caused by the initial trauma. One such skull was found not far from where I work, in the Thames River, close to Hammersmith Bridge, in 1864. This skull has been drilled in the center of the top of the cranium, and the subject of the trepanning was indeed fortunate to escape death (if he did) by surgically inflicted injury to the venous sinus, which runs immediately under this region of the skull.

Another skull was found even longer ago, in 1863, in Bisley, Gloucestershire, England, by Dr. Paine. This skull has evidence of an incomplete operation. A circular groove had been made in the frontal area near the forehead, but the disk of bone had not been removed. Perhaps the patient died during the operation, or possibly the surgeon gave up his task because the bone is rather thick in that area: "Nurse, I said the number 7 flint, please, not the number 6."

There are inscriptions left behind from Sumeria in what is now the southern part of modern Iraq. This was the great civilization that was first started around 5,500 years ago and which subsequently flourished along the Tigris and Euphrates Rivers. Its greatest city was Ur, the birthplace of Abraham. It seems to be one of the first human civilizations to employ writing and tablets, written in cuneiform, and some of these still survive in museums today. Sumerian records mention the treatment of wounds, using oils from pines and fir trees to anoint the cuts, and leaves as makeshift bandages.

Better records exist from ancient Egypt about 1,500 years later. Doctors there were even reported to have performed amputations after serious wounds or accidents. The Edwin Smith Papyrus, now in New York, describes, in 1850 BC, how to deal with injury to the jaw:

If you examine a man having a dislocation of the mandible, should you find his mouth open, and his mouth cannot close for him, you should put your two thumbs upon the ends of the two rami of the mandible inside his mouth, and your fingers under his chin. You should cause them to fall back so that they rest in their places.

War, of course, has been one of the greatest stimuli for the development of trauma medicine. Hippocrates, who treated open wounds with tar—a remarkable forerunner of more recent antiseptic methods—wrote that if a person wanted to become a surgeon he should go to war. Human beings had to wait until we started killing each other in large numbers before trauma techniques were substantially improved.

The first surgeons were almost certainly military. Homer sang of the bold surgeons Machaon and Podalirius, the Roman army certainly had medical officers at least on some campaigns, and wounded crusaders in the 12th century were treated by the Knights Hospitallers. Ambroise Paré (1510–90) was the greatest surgeon of the Renaissance—indeed, one of the greatest surgeons of all time. He pioneered the treatment of gunshot wounds, of fractures of various kinds, and various secondary infections.

From Paris, Paré joined the army in the civil war against the Huguenots. He also campaigned outside France, and his first active service was in the siege of Turin in 1537. Paré writes about being horrified when seeing three soldiers propped up against a wall:

. . . they neither saw, heard, nor spoke and their clothes were still smouldering, burned with gunpowder . . . I was looking at them with pity when an old soldier came up and asked if there was any way to cure them. I said, No, then he went up and cut their throats, gently and without ill will. I told him he was a villain. He answered that he prayed God that someone might do the same for him. . . .

Paré learned that it was not necessary to cauterize, to pour scalding oil on open wounds, and used various balms instead to great effect. He was, indeed, a believer in the Superhuman. Campaigning in Germany, he found one soldier who, on a foraging expedition, was so badly wounded by the local peasants that his colleagues had already dug a grave for him. Here was a test of his skill: "I did the office of physician, apothecary, surgeon, and cook. I dressed him until the end of his case, and God healed him." Indeed, "Je le pansais; Dieu le guarit," was one of his favorite expressions.

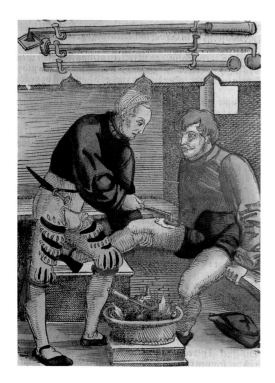

Cautery, the application of heat to a bleeding wound, was a method employed as early as Roman times. This engraving by Wechtlin shows the variety of surgical instruments used in medieval times. At least today cautery is done under anesthetic.

Great British surgeons who treated wounds included William Clowes in Queen Elizabeth I's army and Richard Wiseman, who joined the Dutch Navy and subsequently fought in the Civil War for Charles I. He wrote one of the first textbooks on surgery and has been called "The Father of British Surgery."

The Napoleonic wars gave rise to the invention of the ambulance. Napoleon's chief surgeon was a man called Dominique Jean Larrey. Baron Larrey insisted that the wounded should be transported off the battlefield as quickly as possible, and commandeered horse-drawn carriages to whisk them away to the waiting surgeons. Given that great modern invention, the traffic jam, their response times were probably better than our own ambulance service.

Larrey also invented the first air ambulance—a hot-air balloon. It was greatly to Larrey's credit that he always thought of the welfare of the soldiers. He was determined to feed the wounded properly. On one

The French military surgeon Baron Dominique Jean Larrey amputating an arm during the battle of Hanau in 1813. The patient's leg doesn't look to be in very good shape either.

occasion in Egypt he slaughtered a number of the cavalry's horses to make a soup for the men:

"Have you dared to kill officers' horses simply to feed your wounded?" shouted Napoleon.

"Yes," said Larrey.

"Well," said Bonaparte, "I make you Baron of the Empire."

In the last two centuries, medical advances have come faster and faster (although they have not kept pace with our ability to kill each other). During the American Civil War, infection control was used systematically for the first time, as was chloroform to anesthetize the wounded patients. The First World War saw advances in resuscitation and new methods of blood transfusion, and there were also experiments with the use of seawater to replace the fluids of patients who were suffering from blood loss and shock.

Lessons learned on the battlefield were carried over into the civilian emergency room. The typing and storing of blood, the understanding and treatment of renal failure, vascular surgery, and cardiac surgery, were all

A French soldier receiving treatment in a regimental first aid post during World War 1. One hopes that the phlegmatic assistant on the right did not drop his ash into the wound.

improvised techniques on the front line, and then became successful additions to the medical armory. Surgical techniques for chronic conditions, which have now been refined in the controlled, sterile condition of an operating room, were first tested under canvas in the mayhem of a military hospital, as treatments for injury.

War has changed, just like medicine. As military technology becomes ever more advanced, our armies are refining the so-called surgical strike, in which entire wars can be fought with no more than a handful of casualties. At least, that is what they tell us, and, if it is true, it is only the case for the winning side.

So military surgeons (of the medical kind), bereft of the lines of stretchers outside their deserted field hospitals, need to go elsewhere to practice their skills, and they are returning to the places where such skills are always in demand—the civilian trauma centers, in particular, those located in the war zones of some of the most battle-scarred cities on the

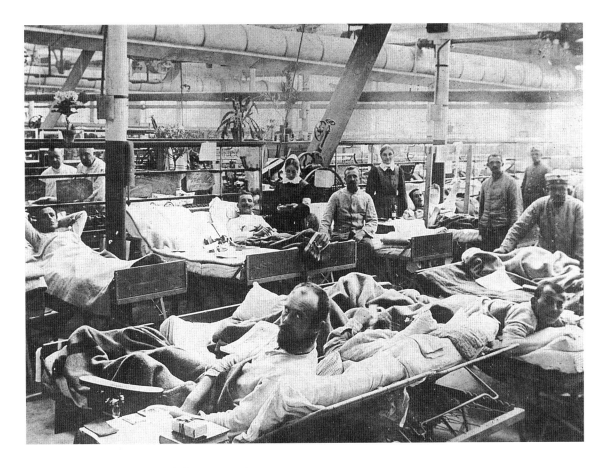

planet. In these grim and dramatic places, trauma is on display twenty-four hours a day, seven days a week. Car wrecks, gunshot wounds, blunt-force traumas, you name it, and the staff of the world's major trauma centers will have seen it—and seen it recently.

A German factory converted into an emergency hospital in 1916.

These days, most new technology and medical techniques are being tested in civilian hospitals and research centers. In the United States, approximately 100,000 people die each year from traumatic injuries. That's the equivalent of two Vietnam conflicts on the streets of America every single year (50,000 American soldiers died in the Vietnam War). Trauma really is a modern epidemic.

If you are under the age of 44, you are more likely to die as a result of a trauma—a car accident, gunshot wound, bad fall—than of cancer or heart disease. In 1997, in the United States, 146,400 people died from injuries or the complications of injuries; more than 32,000 of these were firearm-related. In Britain, gunshot injuries are much less common.

Legislation to control lethal weapons is considered to be justified and humane, and there is no constitutional amendment to serve as a pretext for their widespread deployment.

We are all scared of cancer, AIDS, heart disease, and a hundred other ways of dying, but, with the exception of racecar drivers and professional boxers, very few people think they are likely to get hurt. By sheer numbers, cancer and heart disease lead the list of the causes of death. But trauma mainly affects the young, and, when considering the potential years of life lost in virtually every country around the world, trauma heads the list.

> If you are under the age of forty-four, you are more likely to die as a result of a trauma – a car accident, gunshot wound, bad fall – than of cancer or heart disease.

The trauma industry relies heavily on just one invention—the automobile. Britain suffered its first car-crash fatality more than a century ago, when Henry Lindfield, who was driving to London in 1898, lost control of his car and careered off the road. The cause of the accident was later found to be steering failure. Although Lindfield survived the immediate trauma, his leg had to be amputated later, and he died from the shock.

Today, car collisions take a human life every 12 seconds; 500,000 people die each year and 15 million are injured. In the hundred years since Lindfield's death, 20 million people have perished worldwide as a result of automobile accidents. The latest statistics show that, in the United States in 1998, there were 6,335,000 injuries caused by motor traffic, in which 69,000 pedestrians were hurt and 5,220 killed. During that year, a total of 41,471 people died as a result of a traffic accident.

Thanks to seat-belt safety laws and the introduction of air bags, crumple zones, and shatterproof window glass, injuries of this kind are declining today. Even though twice as many cars are on road, half as many people now die behind the wheel as 30 years ago. Even so, car crashes are the top cause of trauma worldwide; inevitably, this means that the number of people who survive, but sustain serious injury, has significantly increased.

Just what happens to the human body when metal hits metal and internal organs smash is a matter of physics and physiology. A 175-pound man, for example, in a car traveling at just 30 miles per hour, will experience a force nearly 25 times his body weight if that car is stopped dead by another object. Increase the speed to 50 miles per hour, and the catapulting force rises to more than 60 times his body weight.

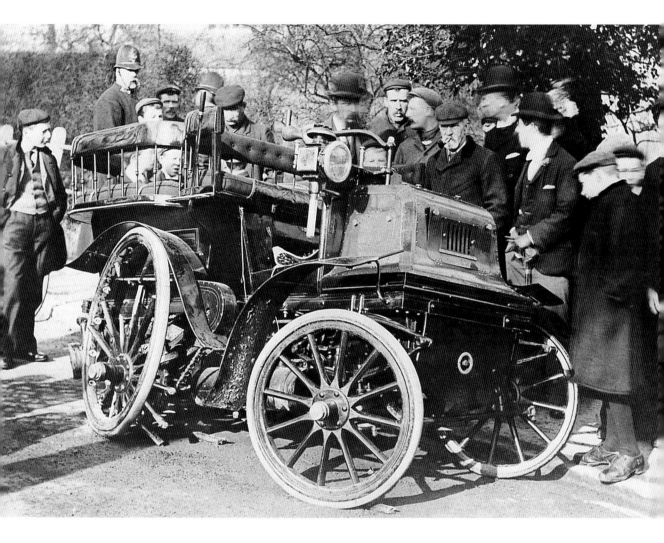

One of the first fatal accidents involving a car in Harrow, north of London, in 1899. Such incidents have always been of extreme interest to spectators.

That force is enough to rip his heart from its mooring, sever blood vessels, and tear delicate membranes. It is enough to detach his retinas, sweeping away his vision in a single jolting moment. The force of the impact is also likely to be savage enough to send fragments of metal and glass flying, ripping the man's skin open, cutting through nerves and severing arteries. And it is enough to destroy the jelly-like network of neurons in his brain, killing outright those that smash against the inside of his skull and drowning others in torrents of blood.

The duration of the impact is measured in milliseconds; within this tiny moment of time a human body can be torn apart and a life irrevocably changed.

Peter Blaud vividly remembers the aftermath of his accident. He remembers seeing his motorcycle lying in the middle of the road. Walking toward it he realized something was not quite right; his right arm seemed to be dangling by its skin, and no matter how hard he tried, he couldn't quite manage to catch his breath. A passing pedestrian told him to sit down, but Peter tried to wave the man away. He didn't need any help, or so he thought.

It was October 21, 1998, and a few minutes earlier Peter Blaud had skidded at high speed into the side of a truck. Peter recalls his train of thought:

"I knew I was hurt. I knew my bike was wrecked. That's why I was looking around, trying to move my bike and trying to clear the accident scene. In my mind I felt fine, but I knew it was hard to breathe, I knew my arm hurt. I was confused, and I didn't know what exactly had gone wrong."

Eventually, several bystanders had to wrestle him to the ground and hold him there. It turned out that Peter's wrist and collarbone were smashed and ten of his ribs were broken. At least one of the ribs had punctured, or, more accurately, had sliced open, a lung. With each breath, his lungs were filling with blood. His spleen was ruptured and his kidneys bruised. So, why did it take more than one able-bodied man to finally subdue him?

Peter Blaud was in shock. Six to eight percent of trauma patients will go into a true state of shock: their hands may become cold and clammy, their skin tinged blue. If the patient has lost a lot of blood, then blood pressure plummets; the brain and other outlying organs are suffering from lack of blood and, therefore, lack of oxygen. The patient is likely to be confused and yet the sudden production of epinephrine, produced to make the heart beat faster, gives the body a sudden burst of energy. This was why Peter was walking around instead of lying on the ground. There are a number of cases in which a trauma victim has walked miles home, often unaware of the seriousness of their injuries.

Epinephrine was essential for battles with big cats and other fast and fierce Stone Age predators, but these defenses, which still spring into action when we are badly injured, were not designed for a world with cars, paramedics, and emergency rooms. They are a patchwork of reactions, some helpful and some distinctively unhelpful, cobbled together from a series of evolutionary compromises. None the less, our Stone Age bodies can tell us a great deal about how to cope with serious trauma.

The ABC of modern emergency medicine is: Airway, Breathing, Circulation. Paramedics or doctors first check that the patient's airway is clear from obstruction. Then they check the inflation of the lungs to ensure that blood is being reoxygenated. Finally, they measure the ability of the heart to pump blood through thousands of miles of blood vessels.

There is no doubt that trauma patients in shock are in serious danger. Peter Blaud was no exception. When he arrived at the hospital, the doctors estimated that his chances of surviving his extensive injuries were less than one in ten. When a patient is shocked, low blood pressure means that oxygen is no longer making its way around the body as quickly and efficiently as it should. If the condition persists, the cells suffocate, and begin to churn out toxic by-products.

Thus it comes as little surprise to learn that one of the first things most paramedics or doctors do for a trauma victim is pump fluids into the body via an intravenous tube. If the patient is shocked, the thinking goes, this will help revival; if the patient is losing a lot of blood, it should stop the onset of shock in a preemptive strike. The blood, and therefore the oxygen, will flow back to the limbs and organs.

Road traffic accidents caused more than 40,000 deaths in the United States during 1998—more than those dying from cancer of the womb.

All this makes it sound as though shock is counterproductive. Are all these automatic physiological reactions "turning against" the body? Why should our blood pressure drop? Why do the blood vessels dilate, so that blood pools in them rather than flows through them? At first, the human body's strategies do not seem to make sense.

After any kind of wound involving blood vessels has been inflicted, platelets stick to the cut edge of the blood vessel. Once a web of fibrin is deposited, a thick clot forms. If a patient is shocked, the blood pressure drops, and the clotting process can be hampered because there is simply less blood, and hence fewer platelets, arriving at the wound site.

Our natural processes start to make more sense when we take into account the fragility of the clotting process. A clot, which forms at the site of a serious wound, is a delicate construction—at least initially. After a time it hardens, and becomes very secure. But, at first, the fibrin mesh can break easily and can also possibly break if the blood pressure is too high. So, it appears that lowering the blood pressure, which the body does automatically, is one way to ensure that clotting has a chance—even though with low blood pressure, the raw materials are thin on the ground, high blood pressure would simply blow the clot, like a surging river sweeping away a dam, and bleeding would continue.

I spent a delightful day in warm May sunshine with Natasha Bondy, the director of the television program this book accompanies, filming from the top of the Cruachan Dam on the west coast of Scotland. The object was to demonstrate how a crack in a dam, or very high water behind it, could increase pressure on the structure and blow the whole construction away.

Rather surprisingly, neither of us could persuade Scottish Power, the authority that owns the dam, to try a realistic experiment. Namely, to allow us to crack one of the walls to see what would happen to the village two miles below. My presentation on camera was done with me standing on the extreme edge of what seemed a very flimsy parapet, with a helicopter hovering just above. A vast amount of good malt whiskey the night before, plus general sleepiness, did not make me feel very secure as the downdraft of the helicopter's rotors really took hold. Warm as the day was, the cold dark water of a loch, holding some ten million gallons, did not look inviting. I was glad that our sound man, the delightful George Hitchens, did not tell me—until after the shot was completed—that the last person who was filmed doing something similar drowned.

OPPOSITE A blood clot magnified more than 3,000 times. The red cells are lining up in formation and are becoming enmeshed in a dense net of fibrin, colored yellow, which gives the clot its extraordinary strength.

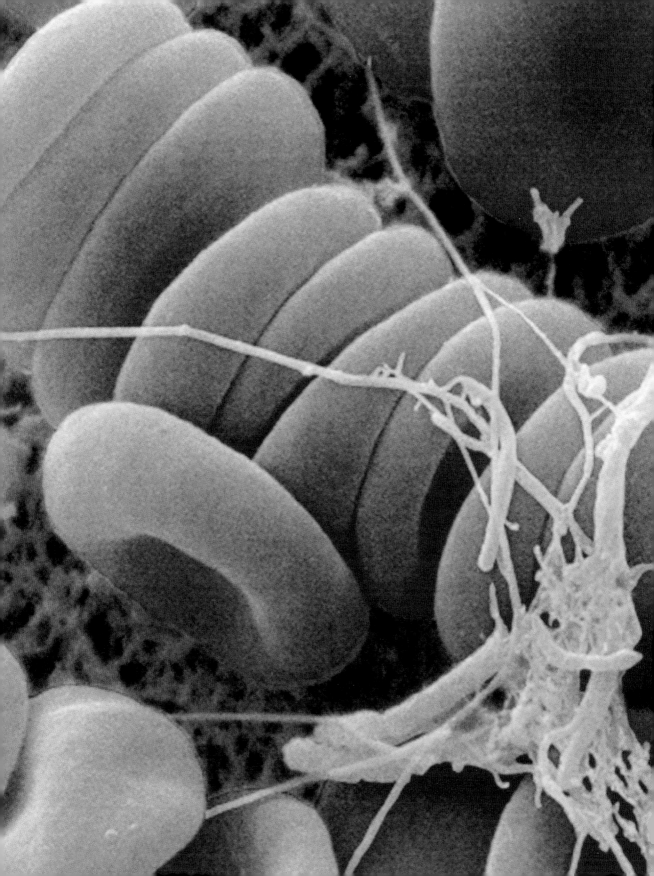

There is not an endless supply of platelets and the other agents that make a clot form satisfactorily. Once the clot is blown, the supply at that moment can be used up. A first clot is precious, a lifesaving plug, which should be protected. Above all, we need to prevent further blood loss after serious trauma, and the first clot is the best weapon we have to make sure this happens.

In Vietnam, the US military was proud of the fact that injured soldiers were evacuated to the MASH units very quickly. The platoons on the front line could call on fast and effective helicopter support. These soldiers had the usual gamut of gunshot, shrapnel, and explosion injuries, which, naturally, meant that blood loss is a prime consideration.

After the paramedics had loaded the bodies into the helicopters, the first thing they did was to put in an intravenous line to try to return their patients' blood pressure to normal. This is standard paramedic procedure—raise the pressure and save the patient. And these paramedics were well trained; they managed to do this more quickly than usual. But the survival rate of these soldiers turned out to be far from good.

Now we understand why. Intravenous fluids would have almost certainly increased the blood pressure to the point where any half-formed clots would have blown, especially since the soldiers were treated so quickly; indeed, they were, it seems, treated too efficiently.

There is a delicate balancing act to perform and when we think about the hard "choices" a traumatized body has to "make," the mechanics of shock start to make more sense. If the injuries are so severe that bleeding cannot be stanched by a simple physiological "bandage"— that is, by building up some clots at the injury site—the body's priorities have to change. Its first priority must be to save the most vital organs. Therefore, once the initial epinephrine effects are over, the body will deprive the less important or more distant cells of the muscles, skin, and intestines. In order to rescue the heart, brain, and lungs, blood flow is directed to them. Rather than allow all of the blood to flow out of gaping wounds, the heart sequesters and preserves it.

Shock, therefore, until now largely seen as a symptom to be treated, is starting to be recognized as one of the body's key strategies for survival. True, it may not be the most effective of strategies, but the body acts in its own best interest. That is why the use of intravenous fluids, rather than saving lives, may actually be detrimental, even deadly.

This view of the use of intravenous fluids is unorthodox. Paramedics still give extra fluid to trauma patients as a matter of course. But there are doctors in the UK who say they would sue if a paramedic attempted to give them fluid after an accident. And in the United States, one of the most highly respected trauma surgeons, Professor Ken Mattox, conducted studies of his own. He vociferously criticizes the standard practice in which fluids are given at the scene of the accident. He is convinced that many more patients could be saved, and that increasing blood pressure can only end in trouble and, sometimes, tragedy.

Shocked patients are always cold. As the shock mechanism lowers the blood pressure, less oxygen reaches the liver, the organ that is responsible for much of the energy production in the body. With the reducing blood supply, the liver is less able to produce heat and our core body temperature begins to fall.

This process is extraordinarily quick and powerful, and virtually impossible to replicate artificially. To demonstrate how difficult it is to reproduce hypothermia, to become hypothermic, I traveled with the film crew, 200 miles north of the Arctic Circle, into the middle of the long Arctic winter. Right at the northernmost part of Sweden the people are Laplanders; their language is a form of Finnish. The Sami people who live there used to be nomadic and, to this day, they still live mostly on the produce of deer-farming and forestry. More recently, with the finding of many minerals there, the area has become rich.

You have to be tough to live there. We were there on a relatively warm morning when the temperature was 14°F (–10° Celsius). In Kiranu, where we filmed, there is a huge igloo with some 40 rooms, large halls, and a bar. The washing facilities are primitive—in an adjacent hut, across packed ice, some 350 feet away. This is what these people call a hotel. Each year it is rebuilt with thousands of blocks of ice taken from the nearby river. The whole place—beds, chairs, tables, chandeliers, bar—are carved from ice. A few rather smelly deerskins cover where one sits, but, believe me, it is cold.

In order to demonstrate how difficult it is to get hypothermic, and how well the unshocked body can preserve its temperature, the lovely Natasha made me stand barefoot on the ice, thermometer in mouth, dressed only in my underpants. At least, the BBC had given me some clean silk ones for the purpose.

After about an hour of filming—at six in the morning—and another heavy evening beforehand drinking schnapps out of glasses carved from ice, I was feeling far from well. But even when the cold became really painful, my body temperature measured in my mouth was still about 94° Fahrenheit, 4.6 degrees below normal. But it took a very long time to start to feel warm again, and there was an extraordinary feeling of fatigue on the journey back to London that evening. What, of course, is so interesting by comparison is that shock reduces core temperature, because of metabolic changes, and cooling sets in within minutes.

A drop in core body temperature has some troublesome implications. Paramedics usually throw blankets on the patient at the scene of an accident. This attempt to warm them up can also aid clotting. Blood clotting is "designed" to occur most effectively at normal body temperature—the last thing that would be evolutionarily successful would be if clotting became easier in severe cold. If the wounds are too serious for an initial clot to form and the core temperature has fallen just a few degrees, more blood will be lost. This is the next stage in a vicious cycle which leads to blood pressure falling again, and the core temperature dropping even further.

Another part of the vicious circle is metabolic. Metabolism will eventually slow after an initial injury. But when the body cools, metabolic reactions are slowed still further. The chemical reactions which take place in the body and produce heat also work best at normal body temperature. One example of a spin-off effect of this is that a number of less essential functions are slowed or halted. The process whereby the bone marrow produces red blood cells is slowed down. This conserves energy for more immediately important matters, but it also means that a really cooled injured person may take longer than average to replace blood cells after an accident. There are also doubts about the effectiveness of white blood cells to deal with infection at low temperatures, a problem if the patient has open wounds.

Not surprisingly, then, all given knowledge would argue that keeping an injury victim really warm is best. Paradoxically, this may be completely wrong. Recent research suggests that our traditional fear of the cold is misplaced. Cold may be a lifesaver, and the blankets may be doing more harm than good.

The brain is an extremely delicate organ and, deprived of a constant supply of oxygenated blood, brain cells very soon begin to die. Just a few minutes after the blood supply to the human brain is cut off, brain damage will usually set in. This process takes a matter of minutes—five minutes, on average. Serious brain damage cannot be repaired. Moreover, once a few brain cells are damaged, there is a cascade effect.

As we shall see from the experiments of Dr. Saffer, when brain cells die, adjacent undamaged cells are badly affected by the chemical substances

OPPOSITE Here I am standing barefoot on ice, clad only in some rather nice silk underwear provided by the BBC. In the experiment my temperature dropped by only 4.6° in 4 hours but I felt ill and tired for 12 hours afterward.

released by the damaged cells. This collapsing domino effect is most frequently seen with a head injury. Damage to one part of the brain, from a blow to the head, may cause the death of cells for some distance around as dying cells send out their poisoned message. This is one of the key battlegrounds of trauma medicine.

So the brains of people who have lost a significant amount of blood, particularly those with head injury, or those who have suffered cardiac arrest so their blood is no longer pumping, are at very great risk. It turns out that simple cooling may be very effective in preventing much of this damage and limiting its extent.

There are occasional stories of people who have fallen into icy water, suffered cardiac arrest, stayed under the ice for a substantial length of time, and survived with their brain intact and fully functional. The most extraordinary of these cases involved a 29-year-old Swedish doctor, Anna Bågenholm, who was skiing in northern Norway when she fell into a hole in the middle of a frozen river. In the current, and in the darkness and shock which followed, she could not find her way back to the surface.

It seems, however, that she was lucky enough to find an air pocket below the ice. She struggled for 40 minutes before losing consciousness, when presumably she was totally submerged. At least another 40 minutes elapsed before the rescue team arrived. On reaching the hospital, after she had been flown there, her body temperature was recorded at 55°F (13.7°C), which is 43°F (23°C) below normal. When Anna was in the emergency room the doctors had recorded the lowest body temperature ever measured in a human survivor.

Anna had no heartbeat or blood circulation. She was not breathing, and her pupils were widely dilated and unresponsive to light. In fact, she was clinically dead, and stayed that way for a full three hours.

She was placed on a cardiopulmonary bypass machine, and her blood was gently warmed before it was returned to her body. After an hour, her heart started beating again. Her resuscitation took a total of nine hours, after which she needed intensive care for another month, followed by months of rehabilitation. The doctor who treated Anna believed that her survival depended on the fact that for the first 40 minutes she was trapped in the air pocket in flowing water, so she had been taking in oxygen as her body slowly cooled down. She was also extremely physically fit.

When she came around from her sedation Anna was paralyzed from the neck down. She had suffered minor brain damage, but it was not

serious enough to be permanent. She eventually regained the feeling in her limbs and, after intensive rehabilitation, was skiing the following Christmas.

Cooling and hypothermia have intrigued trauma doctors for hundreds of years. Hippocrates, in his discussions regarding head injuries, said that "a man will survive longer in winter than in summer, whatever be the part of the head in which the wound is situated."

The screen in the background shows Anna Bågenholm's rescue from the ice-hole. At this press conference, months later, she still shows the results of the physical and mental injuries she experienced. She survived her ordeal despite being the closest a human has ever been to suspended animation.

In both the Second World War and in Vietnam, physicians believed that cooled trauma patients would survive longer. French military surgeons in Indochina actively cooled combat victims, although it is not known whether the survival rate improved. Hypothermia has been used in surgery for many years.

In Russia, where there are not enough heart-lung machines to go around, instead of artificially maintaining the circulation and breathing, surgeons pack the body with ice. This works on exactly the same principle that saved Anna Bågenholm. The patients cool down and go into a kind of hibernation. Metabolic activity decreases, they need less oxygen, and apparently no tissue and brain damage occurs even though the heart has stopped. Muscle relaxants are used to prevent the patient from shivering, which is the automatic response whenever the body is cooled.

Cooling, because it slows down chemical reactions, is a way of suspending time or at least slowing time to a crawl. Therapeutic cooling is, practically speaking, copying a process already practiced by other species. Turtles, for example, hibernate for six months of every year. Submerged in cold water, their heart slows down until it beats just once every ten minutes. The turtles do not breathe for the entire six months because their blood is completely devoid of oxygen. Yet, every spring, they slowly wake up, as if from a good night's sleep, and resume their life, their brain intact.

So we know that cold can be tremendously helpful. But how cold is cold? Inducing severe hypothermia artificially can be extremely dangerous, as Russian experiments on dogs have confirmed. Experimenters cooled the dogs by packing them in ice, stopped their hearts, and then attempted to revive them. The initial results were extraordinary. These warm-blooded mammals were able to survive one, and sometimes two hours of clinical death. The colder the Russians lowered the dogs' core temperature, the longer the length of time the dogs survived. And when their hearts were jump-started and their bodies warmed up, it appeared they had come through the trial unscathed, with no brain damage. Their eyes were bright and their tails were wagging as furiously as ever.

But later the researchers realized that the dogs had not survived unscathed; in fact, they were suffering severe side effects. They found out that the dogs' hearts had begun fibrillating. Fibrillation is when the heart muscle twitches uncontrollably, which is potentially lethal. Other internal organs had also been damaged. Extreme cold, then, was not an answer.

These experiments were based on a long-held view of why brain damage is prevented in the first place. In extremely cool conditions, the theory was, the body's metabolic rate is radically reduced, and the cells' metabolic processes, therefore, are slowed down. Because metabolic time is slowed down, lack of oxygen, or hypoxia, does not cause as much damage. In other words, as the brain is put into hibernation, the process of cell death is slowed to a crawl. This was why the Russians believed that colder was better.

But this seemingly obvious deduction was not, in fact, the correct explanation. Peter Saffer of the University of Pittsburg, has, in his own words, devoted his life to "cheating death." His investigations concern the processes behind brain damage. Saffer realized that hypoxia was not the sole cause of the irreparable brain damage; he researched the chemical processes that begin when hypoxia sets in and discovered that, in effect, the brain self-destructs.

The damage is not caused directly by the lack of oxygen. When the brain senses that there is a lack of oxygen, a chemical cascade is released. These chemicals attack the brain even before hypoxia itself has any chance to cause damage. Conversely, the reason that a brain has survived intact in cold conditions is not simply because the metabolic rate has slowed, it is because the chemical cascade has not been initiated.

Cells can die in one of two ways. They can die by necrosis, which most commonly is the result of cutting off the blood supply. When necrosis occurs usually many cells perish together and they are effectively murdered by an outside influence. The other cause of cell death is by apoptosis—death caused by the individual cell itself, which is effectively a form of suicide. The DNA inside the cell nucleus breaks up, the cell then shrinks and withers away, and it is then eaten by its neighbors. Apoptosis is genetically controlled and the signals to switch it on can occur in a number of ways. Cells in the brain, which are starved of oxygen, may flip on this genetic cell switch prematurely around the damaged area and the cells self-destruct.

Experiments have found that rats whose body temperature had dropped by just four or five degrees, survive trauma without any discernible brain damage. Similarly, in humans, lowering the body temperature by just a few degrees may prevent this chemical cascade from gaining a foothold.

These ideas might just revolutionize the way we handle trauma. It means that we can protect patients from brain damage by cooling them down just a few degrees, thus avoiding the fate of the supercooled Russian dogs. Cooling, used as a kind of treatment to protect the brain, is finding its way into other hospital departments.

Intuitively, it seems right that premature babies are kept warm and snug in their incubators. But these babies are particularly susceptible to brain damage. The lungs of premature babies are only half-developed, and absorbing oxygen is a real struggle. With the introduction of respirators, which effectively take over the task of breathing for the baby, doctors are able to save the lives of premature babies born after just 28 weeks. Now, with the use of drugs that increase the oxygen uptake, we can extent that limit to 25, 24, even 23 weeks.

But a 23-week baby is extremely delicate, and only ten percent of babies born at this stage survive. Of those that do, half will suffer brain damage and grow up to be severely disabled. The theory seems to hold true in these cases—that when the level of oxygen in the bloodstream is low, chemical messengers tell the cells to initiate the process of apoptosis. And the brain of a baby born at 23 or 24 weeks has a lot of development still to do. Unexpected cell death can wreak havoc.

The brains of such immature babies look completely smooth, without the indentations and contours of an adult brain. The baby has yet to make the millions of connections necessary for normal development. This period of development is a critical time, and the reason why brain damage is a major threat for babies born much too early.

At Hammersmith Hospital in London, Professor David Edwards, one of Great Britain's leading neonatal specialists, uses cooling to save the brains of small babies. David Edwards has done something counterintuitive and revolutionary. He cools the babies, using an extraordinary instrument—a broken hair dryer. This amazing technique, potentially useful in any third world country, has saved the lives of babies expected to die. It has also saved other babies—who, based on the severity of their problems at birth, would almost certainly have been brain-damaged; these babies have turned out to be completely healthy and normal.

Kristin Gillespie was a baby helped by cold. She was born at 23 weeks, and was immediately pronounced dead by the doctor. She was not breathing, her eyes were fused shut and there was no discernible heartbeat. Her mother, Johanna, remembers holding her lifeless body, which felt so cold. Then the nurse took Kristin away and prepared to send her to the morgue.

But the morgue was busy that day, and Kristin's body lay unattended in a hospital corridor for an hour and a half. When a nurse finally came to take her away, she got a shock. This supposedly dead baby was breathing.

The obstetrician, Charles Lampley, was dumbfounded. He woke up Johanna with the news that her daughter was not stillborn. Kristin was alive. But he also told her the chances of Kristin staying alive were very slim indeed. Her breathing was very shallow and her heartbeat irregular.

When Kristin was taken to the nursery, her body temperature was low—she was hypothermic. She was breathing only three or four times a minute. Dr. Lampley expected her to live for only a few moments. But during the afternoon she amazed the medical team and her parents by staying alive and gradually getting stronger. Although Kristin was still living, the doctors warned her parents that if she did survive she would be profoundly disabled. She had been deprived of oxygen for about four hours; they gave her a five percent chance of survival, and, almost certainly in such a brain-damaged state as to be a vegetable.

Kristin's parents were determined not to give up on her. She was transferred to a hospital with an intensive neonatal care unit. It would be months before they would know if she was healthy.

Today, Kristin Gillespie is a completely healthy, bright, active 11-year-old, who probably owes her extraordinary recovery to the cold hospital corridor. When Kristin was hypothermic, her cells did not press the self-destruct button, despite the lack of oxygen in her blood. She was protected from the chemical cascade that ends in brain damage.

These new theories and unfamiliar methods will fashion a new style of trauma medicine. It is all about tactics. Some of the body's natural defenses are helpful. We should allow the blood to clot and not force the blood pressure up too high. Cooling may protect the brain, but there is trade-off to be made—blood cannot clot easily at low temperatures. This is a complex contest, which needs to be played with the utmost care and intelligence, not with the cut and thrust of *TV* emergency room dramas.

The debate over intravenous lines demonstrates that it is not a good idea to push for high-tech treatments on a trauma simply out of medical machismo; the trauma is likely to push back. And the promise of therapeutic cooling shows that the intuitive response, keeping a patient warm, is not necessarily the right one. These are lessons that trauma medicine has been slow to absorb. It forces physicians and surgeons to overturn long-held, almost sacrosanct, beliefs.

A humbling experience it may be, but a necessary one, especially when we are trying to care for a bruised and battered Stone Age body in a world designed for the Superhuman.

The banana-import industry was considered to be relatively risk-free until one July morning in Los Angeles, when the Pan American Banana Company's warehouse exploded. Several tons of ethylene, a highly flammable chemical used to speed the ripening of fruit, had ignited, sending produce—and several workers—flying through the air.

The charred remains of bananas, onions, pears, and papayas were strewn throughout the warehouse and an adjacent alley. The fire that followed the blast took more than an hour to put out. The flames reached a height of more than 100 feet and the plume of dusky smoke was seen for miles around. Inside the building, one man had died as a result of the explosion. Five other workers were injured, including Guillermo, who was immediately transported to the LAC+USC Medical Center, luckily into the capable hands of Dr. Demetrios Demetriades.

Dr. Demetriades is in charge of Trauma and Critical Care at the largest trauma unit in Los Angeles. He has vast experience in all conceivable kinds of trauma and is not discouraged by the most serious and seemingly hopeless case. From the paramedics' report alone, Demetriades knew Guillermo was in critical condition. And when the patient arrived at the unit, that report proved to be a wild understatement. The man had burns on over 80 percent of his body, a large skull fracture, two blood clots in the brain, extensive fractures of his facial bones, injuries to his lungs from inhaling fire and smoke, and severe intra-abdominal bleeding that was the result of what is known as the most severe form of liver injury.

Within the first few hours of treatment, he received about 38 pints (18 liters) of blood—three or four times his blood volume. As we have seen, blood clotting is often a problem after injury, and it is much more difficult after a massive blood transfusion like this. Stored blood can only be kept in a liquid form before transfusion by adding compounds that reduce its ability to coagulate. Even with antidotes to these compounds, transfused blood, with its low platelet concentration and abnormal chemistry, never clots as efficiently as it should.

It quickly became evident that Guillermo's injuries were almost too much for his body to bear. Demetriades realized almost immediately that trying to deal with all of the problems surgically, trying to make this man whole again in one fell swoop, was doomed to failure. Dr. Demetriades was convinced that normal trauma procedure, which would involve attempting to fix burns, wounds, and internal injuries quickly and simultaneously, would be fatal:

"He would have died on the table, and I can say this with almost absolute certainty. The injuries were catastrophic, and the bleeding was massive and uncontrollable. I'm certain he would have died."

This prognosis gave Demetriades an extremely difficult decision to make. He decided to do something that is nearly unheard-of in surgical circles, he backed away. That is not to say he gave up; he wanted to give Guillermo's body a chance to recover in its own time. The decision meant embarking on a particularly dangerous high-wire act that is only performed in the most serious of trauma cases.

Demetriades takes the view that the human body has finite physiological reserves. It can take huge injuries but there comes a point when the damage is so profound that these reserves are quickly exhausted and the body tips over the edge. He concluded, therefore, that the best way to avoid squandering those reserves was to control the catastrophic injuries in need of immediate attention, and leave the rest for later.

This *laissez-faire* approach to the treatment of major trauma is referred to as damage control. In truth, it is very much a hands-on process, a detailed and delicate game of chess that is only played in the most dire situations. Damage control is only for those bodies that are spiraling out of control.

It is not difficult to detect a body in physiological free fall. So much blood is lost that there are not enough red cells and fibrin to form a clot. It is a body that has been in shock for so long that its temperature is rapidly plummeting below the levels at which cells can function. Those cells cannot expel carbon dioxide, and so are beginning to drown in a sea of acid.

… blood clotting is often a problem after injury … Stored blood can only be kept in a liquid form before transfusion by adding compounds that reduce its ability to coagulate.

Knowing that long, drawn-out surgery would be more than Guillermo's body could bear, once the neurosurgeons had drained the largest pool of blood from his brain in an attempt to stop it from swelling, the first and only order of business in the operating room that morning was to control the bleeding from Guillermo's severely lacerated liver.

There was not going to be any fancy surgical footwork here; indeed, there wasn't even going to be any real repair. Instead, Dr. Demetriades put in just a few stitches, deep down inside the liver, to pull its severed parts together. Then he applied a tissue glue, and over the tissue glue applied an absorbable mesh, and over the mesh he layered gauze to pack the liver in tightly. This is not elegant surgery—it is simply quick, inflicting the minimum amount of stress on an already traumatized body.

Pressure in the abdomen was likely to build up as the organs and tissues swelled. For this reason, and because he hoped to be able to get back into the abdomen at a later date to finish the job, Demetriades did not close up Guillermo's lacerations. Instead, he and his team slapped a layer of plastic—a sort of prosthetic skin—over the open wound, and wheeled him off to intensive care. Then, they waited, as Guillermo's body valiantly fought death.

For days he lay there, swaddled in bandages and tubes barely recognizable amid the poles and monitors. On the third day, when Guillermo appeared to have stabilized—although well before he regained consciousness—Demetriades brought him back to the operating room. There, he lifted up the sheet of plastic. Underneath, he found the liver tightly packed and beginning to clot. There was some minor oozing and bleeding, which he controlled with tissue glue. Demetriades put in some drains to get rid of the excess blood and then tried to close up. Guillermo's bruised and swollen bowel was in the way, however, so instead, they decided to graft in a new material made from pig bowel to cover the gap.

Allowing the swelling to take place unimpeded is a crucial part of damage control. Leaving the abdomen open has great dangers, for there is a greatly increased chance of peritonitis, which is a potentially fatal inflammatory condition. The use of temporary skin substitutes allows this risk to be controlled.

When Guillermo left the operating room, Dr. Demetriades was pleased but unwilling to declare victory. The bleeding was under control, but now the primary concern was the severe head injury and the brain swelling. The fracture of the cervical spine came in a close second. Third was the lung damage caused by the inhalation of toxic fumes.

At the time, Dr. Demetriades realized that Guillermo would be in intensive care for weeks, maybe months, but remained cautiously optimistic. That optimism turned out to be justified. Just weeks after his accident, Guillermo was sent to a rehabilitation center. He was weak, but on the way to recovery. There is a good chance that, with good support and physical therapy, he will return to full health.

Dr. Demetriades sees damage control surgery as the state of the art in trauma medicine, but it is not a technique for the fainthearted. And yet, he says, a lot of surgeons think damage-control operations are for the inexperienced. The surgical mindset insists that you really need to fix this liver, whatever the implications. This is not an uncommon view. Doctors feel better doing something rather than nothing.

I think, unconsciously, they also feel that out of sight is out of mind, that closing up surgical wounds and making everything neat and tidy is preferable to messy unfinished damage limitation. But this mindset does not admit the possibility that the human body may do a better job of repairing itself than a surgeon's hands can. It does not recognize the skill, the experience, and the nerve that it takes to let go of the controls and let the body fly on its own. Dr. Demetriades says: "After a while, you realize that it's not in the patient's best interest to try and do the definitive anatomical repair."

Sometimes the less you do, the better the result.

Mortality from trauma can come during one of three distinct phases. The first phase consists of those who die in the field. These are the people who are dead before an emergency medical team can reach them. They account for 50 percent of mortality statistics.

The second stage when death can come may be in the first 24 to 48 hours after the event. Patients survive the race to the emergency room, but the doctors are unable to reverse the devastating processes that occurred during the injury, such as massive brain damage or injuries to the heart and other critical organs. These patients account for about 30 percent of trauma deaths.

Even those who manage to survive the first two or three days are not yet in the clear. Some 20 percent of all trauma deaths occur several weeks after the injury. These are the people who, having survived the initial injury, emergency surgery, and other procedures, get past the crucial first day, and may regain consciousness; although still dependent on machines to keep them alive, they seem to be regaining strength. But then the dominoes start to fall one by one and the major organs of the body—the liver, kidneys, lungs, and brain—begin to fail. This overwhelmingly destructive process, responsible for thousands of deaths each year, is called "total organ failure."

This fierce and terrifying condition threatened both Peter Blaud and Jennifer Vaughn. Both were badly injured in highway accidents; both were battered and bruised. The stage was set for multiple organ failure, but could it be avoided?

Nicholas Vedder is a reconstructive surgeon at Harborview Hospital in Seattle and Associate Professor of Surgery at Washington University. His research work is rooted in the view that some modern medical techniques can compromise the body's immune system; he believes total organ failure is the result of treating trauma patients too well and too quickly.

There is no precedent for modern trauma techniques in our body's evolutionary history. Before paramedics and emergency rooms came into existence, trauma victims would have either come out of shock gradually, or they would have died on the spot. After the saber-toothed tiger had moved on to its next victim, the lost blood and fluids would not have been replaced. Now, as previously mentioned, paramedics at the scene of an accident artificially restore the body's fluids. Dr. Vedder believes the body overreacts to this process, sending cellular signals that eventually lead to disaster.

Oxygen-starved cells, once resuscitated, begin to complain loudly about the damage they have sustained and ask for help cleaning up. The immune system responds to that chemical SOS by sending floods of white blood cells, called neutrophils, into the bloodstream. Their job is to find the sites where the damage is worst and remove all the dead and dying cells they can find.

As soon as the neutrophils are activated, they move to the lining of the blood vessels, start rolling along it, and eventually stick to the inside of the blood vessel at the site of injury. When they have reached their destination, they release a number of chemicals. Some of these chemicals create gaps in the walls of the tiniest blood vessels—the capillaries—that allow the white cells to slip between the cells of the vessel walls into the tissue itself. Leaking vessels, together with the neutrophils which seep out, cause inflammation, which is manifested as redness, heat, and swelling at the site of a cut or bruise.

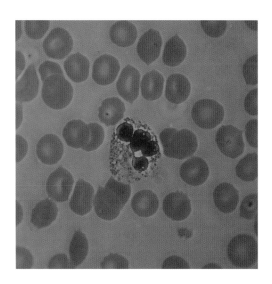

The neutrophil in the center of this picture of blood cells looks innocent and harmless—and yet it can be responsible for clogging up important vessels after injury.

For injuries where the damage is limited, this process works perfectly. But with serious trauma the body's defenses have been breached to such an extent that the small trickle of white cells quickly turns into a surging river, "demanding" so many neutrophils to be called to so many sites that they literally clog up the bloodstream. So much fluid seeps out of the gaps in the leaky vessels that the organs themselves begin to swell. This is why, if you visit a patient in intensive care after a major accident, their bodies are often puffy and swollen —remember pictures of boxers the day after a championship fight?

There is a chance that this process will spin out of control; and if it does, the consequences are severe. Take the lungs, for example. Once the tissue becomes swollen in the lungs, the distance between the air tubes and the blood vessels is widened. The movement of oxygen to the blood vessels decreases. The cells react to this problem

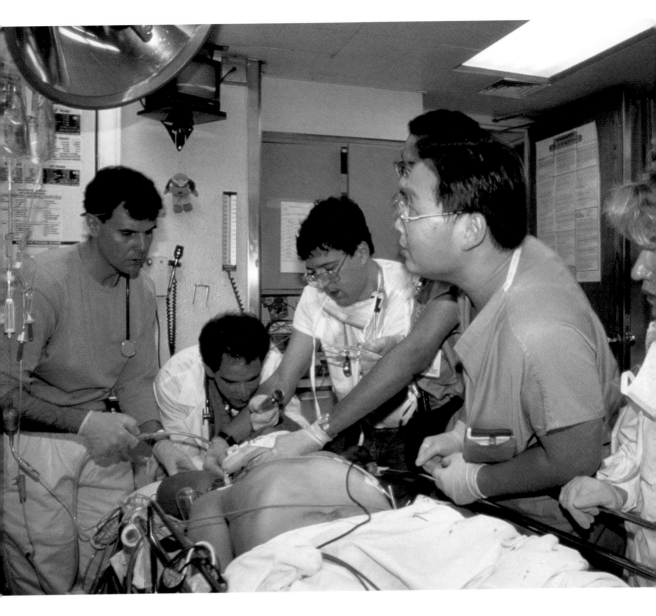

by sending out even more SOS messages to the immune system. More and more neutrophils respond to the call, which blocks blood vessels, damaging the lungs further, and the vicious cycle is underway. The process is similar in the kidney, liver, and other organs. Soon, the damage sustained by the organs is beyond repair. The body turns on itself; this is the condition known as total organ failure—and it is usually fatal.

A key aspect of the treatment of any unconscious accident victim is the need to maintain a clear airway. An anesthetist is putting a tube into the trachea so that the oxygen supply to the lungs is unimpeded.

In order to stop this cycle, we need to halt the white blood cells in their tracks, need to keep the neutrophils from heeding the calls of the cytokines, which are the hormonal messengers. Nicholas Vedder thinks the best way to do that is to stop the neutrophils from taking that first step out of the bloodstream; he wants to prevent them from adhering to the wall of the blood vessel which leads to the release of the vessel-damaging chemicals.

Vedder found that the "stickiness" of the neutrophils is located in a group of molecules that are both on the surface of the blood vessel lining and on the surface of the white blood cell. Rather like Velcro, there are little hooks on one side and wiry fuzz on the other, and the white blood cells stick as a result of this. To stop the sticking, you simply have to stop the hooks from grabbing on to the fuzzy loops. This is easier said than done than it is with Velcro.

Fifteen years ago, Vedder and his colleague John Harlan, a hematologist, began developing a kind of trauma vaccine, which consisted of an artificial antibody that prevents trouble before trouble can gain a foothold. That artificial antibody, called LeukArrest, is a Y-shaped molecule designed specifically to grab hold of a few of the receptors that stud the surface of the neutrophils. With the LeukArrest antibody attached, the receptors can no longer hook on to their complementary loops on the vessel walls. With the Velcro process stopped, the idea is that the chemicals which open up the holes in the vessel walls are not released and damage to the organs is prevented.

The mechanics behind LeukArrest are based on the same principle as the use of aspirin for heart disease. After a heart attack, clots start building up in the blood vessels, and aspirin prevents further clotting from taking place. Experiments have confirmed that LeukArrest does indeed block the ability of the neutrophil to adhere, thus protecting tissues and organs. But experiments in the lab are one thing, clinical trials are another.

Fifteen years passed before LeukArrest could be tested on humans. Why did it take so long? Trials involving trauma patients are fraught with difficulties. Consent needs to be given before a new drug can be tried out on any particular patient. A severely injured and usually unconscious patient is not in the best position to give informed consent, and family members, who could give consent in the victim's place, often do not arrive at the hospital in time. LeukArrest needs to be given as quickly as possible after an accident, because the physiological processes that lead to multiple organ failure begin almost immediately. Once the white cells have begun slipping out of circulation and into the body's tissues, they become much harder to disrupt. Vedder and his colleagues were working against the clock.

When Peter Blaud arrived at Harborview after his motorcycle accident, his injuries were so extensive that his chance of survival was put at just one in ten. His memories of those first few hours are very sketchy, but his main recollection is being very scared:

"My whole body was locked in place. I couldn't move. I couldn't talk. I couldn't breathe because this machine was breathing for me. I still had on a neck brace, so I couldn't move my head. All I could do was move my eyes—I was terrified."

Peter's father, Fred, arrived, and was met at the reception desk by a chaplain. That was fairly indicative of his son's chances of survival. But then Fred Blaud was asked to give his permission for the Harborview doctors to treat Peter with LeukArrest. It was possible, he was told, that his son might be given a placebo. This is a necessary part of any medical trials. But it was also possible he would be given the drug itself, and that the drug might help him pull through his horrendous injuries. Fred did not hesitate in giving his consent.

Peter survived the initial blood loss, but because of the extent of his trauma and the reperfusion injury, he stood a high chance of dying from total organ failure. The doctor making the decisions that night was Ron Maier, surgeon-in-chief at Harborview. Maier remembers sitting by Peter's bedside, trying to decide whether to take him to the operating room and remove his lung in an attempt to save his life. Then, remarkably, the bleeding stopped.

Three days after the accident, Peter's lung function began to improve; two days after that, he was taken off the respirator. And when the double-blind experiment ended, it turned out that he had indeed received the active drug. Maier admits that they cannot prove definitively that the drug was responsible, but Peter had certainly made a remarkable recovery—as had other patients who received LeukArrest.

Jennifer Vaughn was one of those patients. Having been pulled from her mangled car, she was in shock and unconscious. The injuries to her spleen were so bad that it had to be removed. Her liver was bleeding, her pelvis fractured, and blood was pouring into her lungs from the impact of the crash on her chest. She was in mortal danger.

Just minutes after Jennifer's father, Jim, arrived at Harborview, he was asked to give his consent to enroll Jennifer in the LeukArrest trial. Willing to try anything, he readily agreed. Just six days after the accident, Jennifer was able to leave the intensive care unit. The unit housed patients whose injuries were no worse than Jennifer's, but they had been there for weeks, if not months.

However, the promise of the injection that may have saved Peter's and Jennifer's lives did not fare quite as well under the harsh light of statistical analysis. In fact, when the data on 150 patients treated with the antibody at various centers around the United States was examined, it appeared that the treatment had not significantly reduced either the chance of total organ failure, or death.

> ... we need to reinforce the defenses that are helpful, but fight the reactions that are threatening. Sometimes it is not clear which is which.

The data on large-scale trials, like this one, can always reveal a brutal truth. After seeing one or two almost miraculous recoveries, it seems incredible that the drug cannot be doing good. But one or two anecdotal cases, however impressive, could be put down to chance. None the less, Nicholas Vedder remains upbeat about the usefulness of the medication. It did significantly reduce the rate of both heart and lung failure in trauma victims; he thinks the fault may not necessarily lie with LeukArrest itself, but with the frustratingly long amount of time that is needed to administer it.

In tissue that has been without blood for a period of time, and then has blood flow restored to it, the white cells arrive within seconds. Within the first minute, they start to stick and cause injury. If LeukArrest is to be an effective trauma vaccine, it needs to be injected immediately. But since getting informed consent from the patients and their families takes time, often hours.

All new trials of LeukArrest on trauma patients are currently on hold. Instead, the drug is being tested as a potential treatment for strokes, which is similar in effect to reperfusion injury and the inflammatory process which follows it. But Nicholas Vedder remains hopeful that he—and his patients—will get to try it again.

Peter Blaud is back on his motorcycle and has made great progress in regaining his strength. He may never have full lung capacity again, but he is determined to return to his former life as a fanatical jet-skier, which according to Peter, is working wonders as a means of rehabilitation.

Jennifer Vaughn is also back driving a car and generally getting on with her life. Because of her pelvic fractures and the pins that were used to put the bone back together, she had to stay off her feet for almost two months after the accident. That caused so much muscle atrophy that she had to learn to walk all over again, using a walker and then a cane. Today she feels almost back to normal and has even taken up kick boxing, which she says is an excellent form of physical therapy.

So, in the treatment of trauma, as indeed with all medicine, the lines of battle are not fixed; they shift around constantly. We need a fine and subtle understanding of all the processes and reactions; we need to reinforce the defenses that are helpful, and fight the reactions that are threatening. Sometimes it is not clear which is which.

But whatever the complexities of the fight, we need to understand and listen carefully to the body in order to contend effectively. This is what will make us Superhuman, a great leap forward in our ability to save ourselves from the modern world.

Following trauma, there is and always will be a tremendous amount of luck. It is important, say those who dwell in this world of luck and mortality, to remember that trauma can never be cured. People can be saved, accidents can be prevented and guns can be controlled. But there is no drug we can take, no injection we can invent, that will obliterate trauma from the modern world. No matter how Superhuman we become, we will always make mistakes. We will and always have accidents.

> No matter how Superhuman we become, we will always make mistakes—will always have accidents.

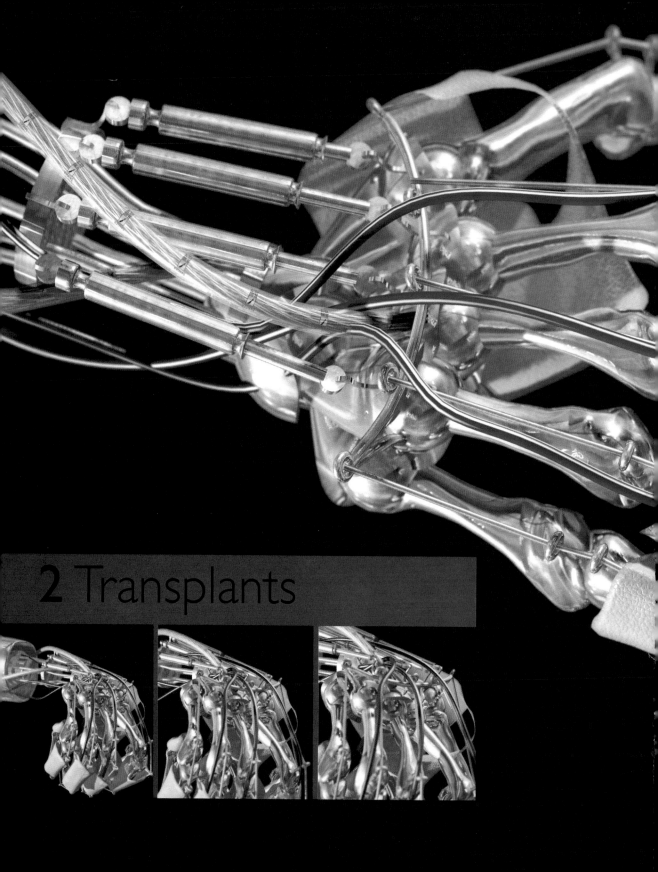

2 Transplants

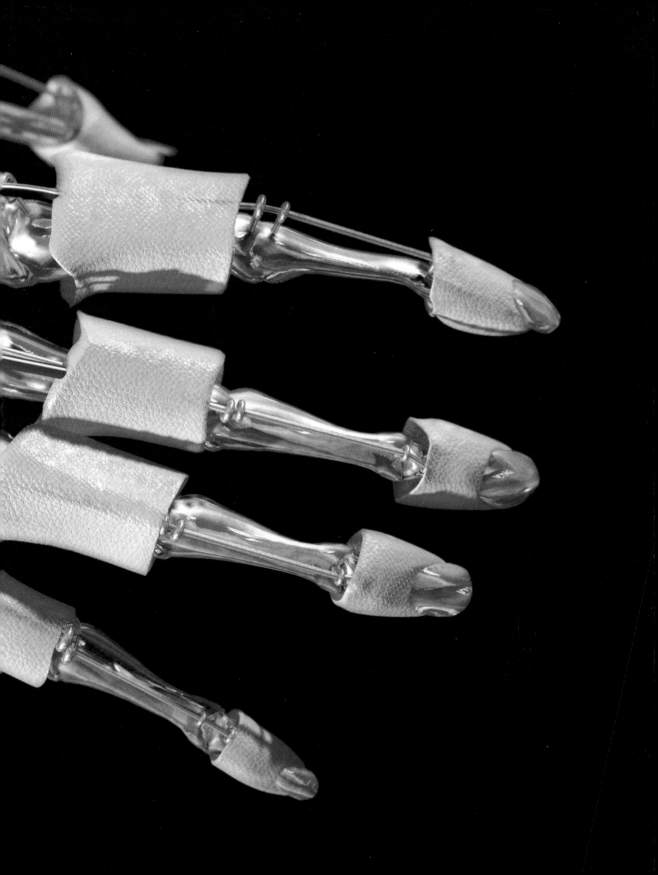

In medicine there is a first time for everything. Fame and fortune comes to those who take risks. Cosmas and Damian, twins from Cilicia, Asia Minor, knew all about risk. These two celebrated physicians, or at least those who promulgated their story, can take the credit for the very first transplant of a human body part.

In AD 280, it is said they were at the height of their fame. A patient came to them, one of his legs riddled with cancer. It was immediately apparent that the leg would have to be amputated. Cosmas and Damian realized this was an ideal opportunity for an experiment. One night, while the man slept, the two intrepid surgeons replaced his cancerous limb with the leg of a recently dead Ethiopian Moor. The patient awoke to find he had a new leg, albeit a black one.

It is not recorded whether the patient was happy with his new leg; in any case, it is extremely doubtful that this daring venture could have been successful. Even if Cosmas and Damian had managed to attach the blood vessels sufficiently well to allow the circulation to resume, even if they had managed to avoid infection and gangrene, it is safe to say that the leg would have begun to rot within days, if not hours. Soon after, the patient would have been among the first to experience rejection, a process in which the immune system refuses to accept alien tissue. To this day, coping with rejection remains the biggest challenge for transplant biology.

The magic of modern technology allows me (right) to observe as twins Cosmas and Damian from Asia Minor examine the patient who will be the recipient of the first-ever transplant of a human body part – a leg – in this recreation of the medieval painting by Francesco Pesellino, 1422–1457.

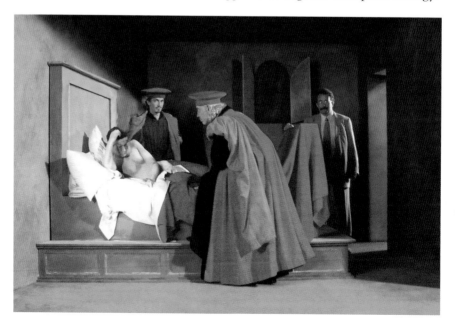

Despite this audacious experiment, or maybe because of it, Cosmas and Damian became the first martyrs of the medical profession. After apparently managing to survive being burned, stoned, crucified, and sawed in half, they were eventually decapitated. Their gruesome deaths ensured a happy ending of sorts; they became the patron saints of medicine, although as role models they leave a little to be desired.

St. Cosmas and St. Damian would be at home with today's trailblazers on the wilder shores of transplant medicine. There are some extraordinary experiments taking place, as we shall see later. We meet a man who may receive a new pair of hands; we meet genetically modified pigs who are specially reared to produce human organs; and there is the man who attempted to mix his immune system with that of a baboon.

Sometimes these bold pioneers may go too far. And, of course, as is so often the case with medical daring, the bravery of these pioneers is less than the bravery of their subjects. There is much debate on the ethicality of research on primates, particularly chimpanzees and baboons. Robert White, one intrepid neurosurgeon in the United States, has performed on monkeys the ultimate experiment in transplant medicine—a head transplant.

Dr. White was, of course, using the monkeys as the ultimate model for humans. It makes more sense to think of his head transplant procedure as a full body transplant. His rationale was that people who are currently completely paralyzed have a diminished life span, because organs tend to fail much more quickly if there is substantial damage to the nervous system. Transplanting an entire body would never, at least for the moment, allow the patient to regain any feeling or movement; the spinal cord would have to be connected, and that, of course, is impossible. However, a quadriplegic with a completely new body would, in theory, have a much increased life expectancy.

The operation involves cutting off the head while keeping a blood flow to the brain and then grafting on a completely new body, carefully connecting the spinal cord, blood vessels, muscles, and tendons. Video footage of the monkeys with new bodies shows they are still conscious after surgery; they blink and lick their lips. It seems extremely unlikely, given the complexities of reconnecting spinal cords, that they will regain any use of their limbs.

After these trials, the surgeon claimed he was ready to attempt the same on humans, and, in fact, he had a number of willing candidates who were apparently prepared to undergo this bizarre procedure. Inevitably, his attempts to persuade the American authorities to allow the trials have met with little success. He may now go to Russia or the Czech Republic instead, where, he believes, the regulations are a little more lax.

However gruesome, the mere fact that the monkeys survived for a while shows that this technical achievement in transplant surgery is astonishing. Surgeons have an incredible ability to attach the most complex organs and limbs. Most of the physical problems of reconnecting and grafting have been overcome.

Transplantation is at its most successful with autotransplantation. Plastic surgeons, for example, can now cut flaps of a patient's own skin, fat, and underlying muscle, perhaps from the buttock, thigh, or chest, carefully preserving the blood vessels after cutting them. The free flap can then be transplanted to other sites in the same person—to fill defects, say, after removal of a cancer affecting part of the head and neck. The use of microsurgery to hook up tiny blood vessels and nerves has revolutionized this particular field of plastic surgery.

But most transplants are needed to replace essential organs, and these have to be obtained from another person. Such transplantation is still not an easy option, thanks to the immune system. Only identical twins will have an identical immune system, and most people do not have a suitable clone to give them a spare kidney, much less a heart. The processes involved in rejection are now understood in far more detail, but the solutions, until very recently, have involved the heavy use of immunosuppressive drugs, which carry a considerable risk in themselves. But even for those patients who have little choice—if a transplanted organ is the only thing that is going to save their life—then there is another problem, one that the brightest medical minds cannot solve: the lack of available organs.

Five men are waiting for a miracle on the Cardiac Care Unit of New York's Mount Sinai Hospital. They have been there for months and they complain bitterly about the food.

Each of the five men is waiting for a new heart, because theirs is failing. The other patients call them the "Pole People" because the men spend their days attached to poles decorated with boxy monitors, bags of saline, and sundry medications. The poles, like the pain and the sense of imminent mortality, rarely leave them; they hover over the men as they play dominoes in the dayroom, and they are dragged along on every lap around the ward. (Seven times around is a mile, the men will explain to you, and each mile logged increases your chances of staying healthy enough to receive the miracle, should it arrive.) Even at night, the poles stand silent guard over their sleeping charges.

When the heart is failing, water, the basis for all life on earth, can mean death. If there is too much water in a body with an inefficient heart pump, the lungs start to fill up and a person can literally drown from within. Therefore, these men are given diuretics, water tablets, and injections to make their kidneys work harder, to dry them out. Their diet is drastically curtailed, so that they savor each bite, even though the hospital food here is so bland. The blandness is partly because so many things, such as salt, that make a normal person's diet tasty, have to be restricted. "I can remember when I was a young second lieutenant and had a roommate, and all we could talk about was girls, how pretty they were and how available this one was, and so on," says 62 year-old Peter Liuzzo, who is a former nurse anesthetist. "Now,

ABOVE Peter Liuzzo (right) awaiting his heart transplant in the hospital. His young companion, Rommel Geradino, had already undergone this life-saving operation.
BELOW Surgeons prepare to transfer a donor heart to the patient.

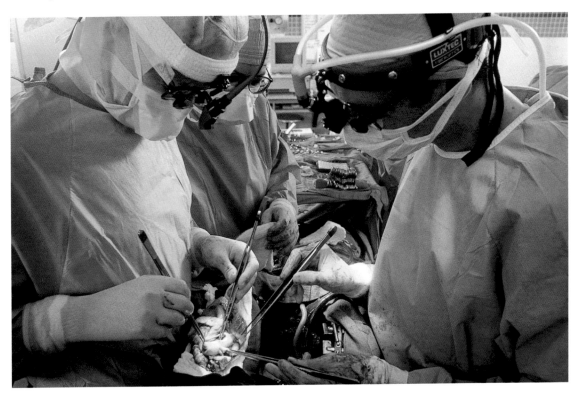

for the second time in my life, I have a roommate, and we all stay awake and talk about food."

They gather together daily to complain about the food. On the really bad days, they might make a visit to Room 318, better known as John's Deli. Here John Weissman, a retired New York firefighter, stockpiles the pizzas his family brings to keep his spirits up. John distributes slices to his cohorts and the occasional friendly nurse, and, on special occasions, he might even whip up baked potatoes and asparagus in a portable oven.

When seat belt use became compulsory in the UK in 1983, the supply of organ donors was halved almost overnight. Ironically, far more people are dying on transplant lists because far more people are surviving car crashes.

There is a feeling of family among the Pole People. Unfortunately, the family is a large one. There are 66,000 names on the organ transplantation lists in the United States today, but the number of donors has stayed relatively steady at 5,906 in 1988, and just 9,913 in 1998. Half of those on the heart transplant list will die before a donor is found.

The situation is no better in the UK. In December 1999, for example, there were a total of 5,396 patients waiting for transplants; 4,691 of them were lining up for a kidney, 212 for a heart, 182 for a lung, 156 for a liver. But, of course, these numbers are really only the tip of a much bigger iceberg. For example, there are probably ten times that number of people, not on a registered waiting list for a heart transplant, who might genuinely benefit from this procedure. There are also many other patients with heart problems who could benefit from a new heart, but who do not meet the stringent requirements necessary to get on the list.

Despite these depressing figures for those needing lifesaving surgery, about 2,000 kidney transplants are performed each year. There are simply not enough kidneys to meet the need. Those who need hearts are in for a longer wait. At least one in ten patients will die on the waiting list, as will far more who never actually get onto it. Half the hearts sent for transplantation are too damaged to use.

Whichever way the numbers are spun, they make grim reading. A major reason is the fact that many organs used in the past came from young people who died quickly and unexpectedly in car accidents. When seat belt use became compulsory in the UK in 1983, the supply of organ donors was halved almost overnight. Ironically, more people are dying on transplant lists because far more people are surviving car crashes.

Those on the list are well aware of the statistics. "We kid around, we have our laughs," says John Weissman. "We'll see a young guy who looks good, and we'll wonder what blood type he is. But it's a hard thing here, because we know somebody has to die for us to live."

Liuzzo remembers clearly the day he arrived at Mount Sinai. "The treetops on March 25th were gray, and there was not a leaf to be seen. And today they are all green. I've watched them go from gray to green," he says, with more than a hint of melancholy, staring out at Central Park. "I watch people. I wonder where he is going, I wonder where she works. We are out of the world right now. Perhaps more and more people will understand that there is a gift to be given."

Twenty years ago the answer to Liuzzo's prayers was plain. We would replace our own organs of blood and tissue with metal and plastic surrogates, finely engineered devices that would make organ waiting lists a thing of the past. These artificial body parts would be even more reliable than the original. We would become bionic. Our prosthetic limbs would be stronger and more dextrous. Our plastic kidneys would filter more finely and last for ever. Our senses would be enhanced; plug-in modules for clearer vision and more acute hearing could be bought over the counter.

Kyle aged 2½, shortly after his cochlear implant, which enabled him to hear for the first time.

There is no transplant list for 2½-year-old Kyle, who is profoundly deaf, the result of a rare congenital defect that left his ear canals undeveloped. Kyle's only way into the hearing world is a cochlear implant, an electronic device with a microphone that provides direct stimulation to the auditory nerve.

There is a risk attached. The surgical incisions go perilously close to the main facial nerves, which, if they are damaged, would leave Kyle's face paralyzed, probably permanently. As she watched her son being wheeled into the operating room, Kyle's mother put the dilemma in blunt terms. "Which would I rather have, a child who could hear, or a child who could smile?"

Fortunately, the surgery went without a hitch. A week later, with the surgeons and Kyle's mother anxiously looking on, the implant was switched on, and Kyle winced, hearing sound for the first time; this wince soon turned to delight.

There's no doubt that mechanical devices—steel and plastic mimics of human body parts—can serve admirably as stand-ins for those parts that are no longer holding up their end of the physiological bargain. There are those in the deaf community who question whether or not an intervention like the cochlear implant is a good procedure. They say that deafness is worth living with and that such implants impose the values of the hearing society on a child who is too young to make a choice. Given the risk, they say, the procedure is not justified. Unfortunately, the operation cannot be deferred until later in life when informed consent is possible; by then, the chance of learning normal speech will be permanently impaired.

Few, however, would question whether it is a technological success. Indeed, if one had any doubt, it would be immediately banished upon watching the look of joy and awe spread over little Kyle's face the first time his implant is switched on and watching the tears spring to his mother's eyes.

The cochlear implant is a device that is now over fifteen years old, and was the first synthetic interface between the brain and the outside world. It is not an infallible solution; in some cases there seems to be no effect, and the patient must have an intact auditory nerve. If the nerve has been damaged, the only alternative is to implant directly into brain stem, or, perhaps, into the auditory cortex. This is undergoing trials, with varying degrees of success, but the technology is still in its infancy.

Developments in material technology are allowing easier integration of the artificial with flesh and blood. Material technology is a relatively new field in human medicine and where I work, at Imperial College in London, it has become of major importance in much research into transplantation and tissue engineering. One of the first substances to be used for human implants was titanium. Titanium, a metal that is comparatively light and strong, does not corrode, irritate the tissues, or cause scarring. It is a material that has been found to be biocompatible.

The applications for its use are very practical. For example, since medieval times—think of those paintings by Pieter Breughel in the 1500s—a wooden artificial leg was strapped into a harness, which, in turn, was strapped rather awkwardly onto the leg. That is still true in many parts of the world. But a hole can now be drilled into what remains of the bone, and a large titanium peg can be implanted, so that the artificial leg can then be clamped securely into the peg.

There are no signs of rejection with titanium. In fact, the bone seems to integrate with it and grow around it, which makes the implants stronger and more secure. Even better, because the titanium is a solid connection between the bone and the prosthetic leg, patients who have tried it say there is much more sensation—a sensory feedback from the leg. They can tell, for example, whether they are walking on grass or walking on a carpet.

The Beggars (1568) by Pieter Breughel the Elder illustrates graphically the primitive artificial limbs of medieval times.

Another tentatively successful story—a story that may be only at the very beginning—is that of retinal implants. Mark Humayun, at John Hopkins Hospital in Maryland, has implanted tiny chips 2 millimeters x 2 millimeters into the eyes of over a dozen blind people, all suffering from conditions such as macular degeneration, which damages the retina but leaves the nerve cells behind intact.

These chips electrically stimulate groups of nerve cells, called ganglia, mimicking the workings of a healthy retina when light falls on it. The chip is connected by a miniature radio link to a tiny camera mounted on a pair of eyeglasses.

Some of the patients in the trial had been blind for 40 years. After the chip had been implanted, it took some time for them to recognize and "see" the signal, but they quickly got used to it, and were bowled over. One emotional 71-year-old man said. "It was just wonderful. It was blue. It was like switching a light on."

At present the resolution of the image is extremely low; some of the patients can just about recognize a large letter or number when these are held right in front of the camera. They see spots of various colors—blue, yellow, and green—but it is not known how to control what colors are seen electrically. However, even if everything does look blue, or green, and despite the fact that we cannot get remotely near the resolution of the human retina, those patients who have tried the implants are not complaining. Even a shadowy, totally blurred pixellated image can make an enormous difference to the life of a person who was formerly totally blind. But I cannot help wondering whether these metal/silicon implants (so inferior to the human retina) are ever going to be a real substitute for retinal cells.

The apparently fast pace, therefore, is still tantalizingly slow for sufferers from these diseases. Nonetheless, progress in technology and miniaturization means that metal, silicon, and plastic devices may occasionally serve admirably as stand-ins for their biological counterparts. In the future, will we be able to pick up artificial parts off the shelf, as we do spare parts for an automobile? What of the Pole People and the interminable waiting lists? Will artificial parts prove to be the solution to the organ donor problem?

The answer, at least for the foreseeable future, is no. Despite beautifully crafted devices, such as the cochlear or retinal implants, there is a growing realization that most organs cannot be replaced by synthetic substitutes. The performance of these implants is far below that of the original. The initial optimism surrounding artificial parts vastly overestimated our abilities as engineers, as well as underestimating the complexity and incredible reliability of our own organs. It also underestimated the complexity of the connections between these devices and the human brain.

There is another reason that major organs are more difficult to replace artificially. If our retinal implant fails, we would not be in mortal danger, but major organ transplants, such as the heart, kidneys, and lungs, need to work perfectly and reliably because our lives depend on them.

There is a recognition that human engineering cannot come close to the real thing. A synthetic organ is worse than a fake Rolex bought from a Bangkok street vendor; it might do the same thing as the more expensive original, but not as well and not for very long. In this sense, the dream of the bionic man has fallen flat, and nothing illustrates this better than the story of the artificial heart.

The heart beats more than 100,000 times a day without a rest. It has a life expectancy of over 70 years, during which time it will beat over 2.5 billion times. It does not require servicing every ten million heartbeats, nor does it need lubricating, cleaning, or antifreeze in the winter. Like any mechanical system, it works better if we take care of it, yet it often survives the insults from thousands of pints of beer, countless cigarettes, and a million greasy french fries.

Even after several heart attacks, the heart, with much of its muscle replaced by scar tissue, can be jump-started and coaxed back into life. This blob of pulsating muscle is a minor miracle; for sheer reliability, it beats any genuine Rolex hands down.

> … in the future, will we be able to pick up artificial parts off the shelf as we do spare parts for an automobile?

The heart is the organ that has always seemed to be the center of the body. Those who dared to open the chest cavity and who pioneered operations on it always had star status, and even today some heart surgeons seem bigger than life. When I was a student I recall one of the early pioneers of open heart surgery, who, on being handed the wrong instrument by the nurse, threw it dramatically across the operating room. Thereafter, instrument after instrument was passed into his grasp, whereupon he growled and threw each one across the room. Soon, the instrument tray was virtually empty, the patient was still anesthetized on the table, and the tearful nurse had been replaced by another. Actually, in spite of the mythology, heart surgery is undoubtably no more difficult than any other form of surgery; much of the responsibility lies with the anesthetist. However, the status of highly paid heart surgeons often went to their head.

Michael DeBakey's heart has been beating for over 91 years. He is a legend; as a leading heart surgeon in Texas, he earned the nickname Texas Tornado for his fiery temper and drill-sergeant style in the operating room. A BBC television program in February 1999 showed archive footage of Dr. DeBakey in the operating room at work. "Don't ever put your hand in front of my line of vision, damn it. I know, but you don't know what I want to do, you see? What do you mean, it's not flushing? No, no, no, no, no. Put your finger over that. See, you're not concentrating, you're watching me." But, in spite of all his dominating behavior, there is no doubt that he became a real pioneer in artificial hearts at an age when most people would be thinking of retiring.

Three decades ago DeBakey and his team engineered a heart that tried to replicate the human heart. It consisted of two separate chambers, corresponding to the left and right ventricles, and a diaphragm in between that pumped the blood through from one compartment to another. The device was bulky, with an external pumping system connected via thick tubes that led into the chest cavity, but at least in the laboratory it seemed to work pretty well.

The year was 1969. Dr. DeBakey implanted the electric heart in seven calves. Four died on the operating table. DeBakey insisted there was a great deal of work to do before the device could be tested on a human being. Two of his colleagues, Domingo Liotta and Denton Cooley, were impatient and decided they were not prepared to wait.

Dr. Liotta and Dr. Cooley found their human guinea pig in Haskell Karp. He had a long history of heart problems, and in the absence of a suitable heart donor, had agreed to be the first man to be implanted with an artificial heart. The two renegade surgeons took one of the prototype hearts from DeBakey's lab. Karp was kept alive on a heart-lung machine while the device was implanted. Dr. Cooley did not ask permission of the hospital or any federal agencies; he did not even tell Karp's wife.

Dr. DeBakey was furious. He was out of town at the time, in Washington, DC, and read about the operation in the morning newspapers. There were pictures of Karp lying in bed after the operation, seemingly in good health. In reality, Karp was not in such a good state; just moments before he had been covered in tubes and surrounded by resuscitation machines. The tubes were removed and the machines were hidden under sheets for the benefit of journalists, who went away and wrote enthusiastically about this miracle of modern technology.

Over the next two days, however, Karp's condition deteriorated. Dr. Cooley and Dr. Liotta resigned themselves to the fact that the device was not doing its job. Karp's wife, Shirley, made an emotional appeal for a human heart; within a day, a heart had been found, and the operation took place, but her husband died shortly after the transplant. That was one fact Dr. Cooley and Dr. Liotta could not hide from the press.

Denton Cooley is unrepentant; he claims he was trying to save the life of the patient, and it is true that Karp would have died anyway without Dr. Cooley's intervention. However, the episode left more than an unpleasant taste in the mouths of the researchers and the public alike. The whole fiasco was an inauspicious start for a program to which great hopes were attached.

Dr. DeBakey never forgave Dr. Cooley, who resigned from the college after the controversy became public. Despite working in the same city, these two accomplished heart surgeons have not talked to each other for more than 30 years.

Dr. DeBakey scaled down his ambition and successfully developed devices that assist weak hearts rather than attempt to replace them. These pumps bypass the left ventricle, which houses the main pump of the heart, and boost the blood flow in a patient whose heart is severely weakened through heart disease. But the big prize of a total artificial heart replacement was still up for grabs. In the 1980s there was a new contender in the race, a young medical engineer named Robert Jarvik.

After several years' work, he decided his Jarvik–7 replacement heart was ready for a test. It worked on the same principle as DeBakey's earlier model, except that the pump was driven with compressed air. It also was smaller and potentially more reliable than the DeBakey model.

Dr. DeBakey did not share Jarvik's confidence in the prototype. He asked Jarvik for experimental data that showed its reliability in animals, but Jarvik ignored him, refusing to hand over the data, and a team led by William De Vries of the University of Utah went ahead with the operation. It must be said that Dr. Barney Clark, the patient, had nothing to lose; his heart was due to give up the ghost at any time, so he gratefully took his place in the medical hall of fame and accepted the Jarvik–7. Not long after the operation, Barney Clark was sitting up and telling the world what it felt like to have an artificial heart, made of metal and plastic, beating away in his chest.

Then, after just 13 days, a valve broke. Dr. De Vries had to open up Barney's chest again and replace half of the artificial heart. The valve in question had never been fully tested in the laboratory. After they fixed the problem, Barney lived for only a few months. His quality of life was poor. He had chronic nosebleeds and needed constant, innumerable tests.

Dr. DeBakey called the experiment a "disaster." Another eminent surgeon suggested that the only thing they had learned from the case of Barney Clark was that the device was not fit to be tested on a human patient. But Robert Jarvik and William De Vries (Dr. De Vries himself on the front cover of *Time* magazine) were famous, and, like Denton Cooley, will have their names engraved in the record books. Their reputations will always be tainted by the accusations that they went too far, too soon, for reasons of ego rather than science.

Meanwhile DeBakey was getting help from a surprising source. He had performed a heart transplant on a NASA engineer, David Saussier, and Saussier became interested in DeBakey's work on the left ventricle assisting pumps, which were called LAVDs. Dr. DeBakey and his grand entourage of technicians went to NASA with their bulky LAVD prototypes, all of which relied on the same principle as the biological heart—using a pulsating diaphragm to pump the blood through. But NASA knew about pumps; the space shuttle depends on pumps designed to move large quantities of liquid through a very small tube.

The NASA engineers had an idea. Why mimick the pulsatile mechanics of the heart? Why not try something completely different? Saussier and a group of NASA engineers designed a pump that worked using an Archimedes screw. A tiny spinning turbine would draw the blood in one end and push it out the other, in a smooth flowing motion. Dr. DeBakey knew this was a breakthrough. It was both substantially smaller, the whole device being roughly the size of a thumb, and simpler, so it would probably be more reliable. The constant, smooth flow had an interesting side effect—the recipient would no longer have a pulse.

There was, however, a problem. The spinning turbines tended to mash up the blood cells and release hemoglobin into the bloodstream, which meant that the recipient would get jaundice and turn yellow. In addition, the patient would, of course, become increasingly anemic, and his or her ability to carry oxygen around the body to vital organs gradually diminished. If you throw water-filled balloons at a fan spinning fast enough, some of them will burst, releasing their contents. The NASA engineers modeled the blood flow using Cray Supercomputers. Eventually they discovered that tiny alterations in the geometry of the turbine blades, as well as increasing the speed of the spin, could minimize the damage and leave the red corpucles unmashed.

The device has met with a good deal of success. At the age of 90, Dr. DeBakey personally

OPPOSITE *The high point of 20th-century bio-engineering—a Jarvik–7 heart made from plastic and aluminum. It worked for thirteen days, beating inside Barney Clark's chest.*

oversaw six operations at which the new LAVD was implanted. The patients, none of whom now have a pulse, have fared reasonably well. In some cases the system is working even better than expected; the device is used in patients with serious and chronic heart disease. In one case, the device has given the patient's existing damaged heart, which is still in place, a chance to recover. Robert Jarvik is snapping at the heels of Dr. DeBakey; he is working on the Jarvik 2000, a device similar to the DeBakey LAVD.

However, there are no winners in the race to produce a complete artificial heart. At least in the immediate future, this race is unlikely to be rerun. The researchers' initial optimism vastly overestimated the engineers' skill and underestimated the complexity and incredible reliability of our own body parts.

Some synthetic substitutes, such as the artificial cochlear implants, have proved their worth, at least to a limited extent. However, the real hope for people who need major organs and other complex body parts remains with the actual thing. The Superhuman will not be packed full of bionic gizmos made of plastic and metal. He or she will be using the best brand available—biological body parts. Why settle for titanium when you have human tissue?

> The Superhuman will not be packed full of bionic gizmos made of plastic and metal. He or she will be using the best brand available—biological body parts. Why settle for titanium when you have real human tissue?

The amputation of James's fingers and toes was the final result of treatment to save his life. James first became ill with a massive infection caused by an invasion of "flesh-eating" bacteria, Group A streptococci. These, as far as we know, made their way into his bloodstream through a tiny cut in his skin.

The infection quickly overwhelmed his body, and his blood pressure plummeted. To this day, James is unable to remember anything that happened to him between February 19, when he was brought into the hospital emergency room in Denver, and March 30, when he finally came around.

His doctors were able to save his organs and brain from permanent damage, but, to do so, they had to use extremely powerful vessel-constricting, or vasopressor, drugs. These drugs shunted blood from his extremities to keep the blood and the oxygen it was carrying circulating around his vital organs, where it was needed most.

The doctors knew when they gave him vasopressors that they were taking a calculated risk. They knew that the drugs were so effective at keeping the abdomen and its organs bathed in blood that his extremities would be starved of oxygen. The longer they kept him on these drugs, the more likely he was to lose some fingers or toes. But without the vasopressors, James was not expected to survive, and so, with his wife's approval, the doctors decided to proceed.

On March 30, 1998, James woke up to find that all his fingers and toes had been amputated.

James says that, had he been conscious, he would have made the same decision. However, he sees no reason why he should accept the consequences. That is the reason why he is hoping to become one of the world's first double hand transplant recipients.

The field of hand transplantation has sprung up because of advances in microsurgery. John Cobbett, a British surgeon, was probably one of the first to use very fine stitch material to transplant a digit. In about 1970, he autotransplanted the big toe of a man onto the stump where his right thumb had been cut off by a machine. And I think that I myself may have been among the first surgeons to be involved with this kind of surgical procedure.

When I was working in Leuven, Belgium, in 1976, I helped Professor Willy Boeckx graft a man's hand back onto his wrist after it had been severed by a machine. I was in Belgium at this time researching attempts to improve microsurgical techniques. Johann, a farm worker, had been threshing corn and his friends, seeing him bleeding and semiconscious, had rushed him and his mangled severed right hand to St. Pieter's Hospital in Leuven.

The operation that followed started at six that evening. It involved Willy and me rejoining several small arteries and veins, and repairing the key nerves. The operation took us 14 hours and at the end of it, the patient had to be taken back to the operating room for another six hours. Additional surgery was needed because some of the vessels had clotted up even though we had thinned the blood with anticoagulants.

Even after this massive operation, Johann needed further surgery by orthopedic specialists to stabilize the wrist joint and properly fix the hand in position. Later on, five more hours of surgery was required to connect the tendons. Johann had a long period of convalescence and months of physical therapy. The hand stayed alive, but Johann could never move his fingers properly, could not grasp an apple, and had constant arthritic pain and discomfort.

Operations like Johann's, of course, are not complicated by the rejection process because they use the patient's own tissues. Apart from being a technically difficult procedure, a donor hand transplant would really be no mean feat. Previous attempts to transplant hands on monkeys failed because the monkeys violently rejected the donor appendages, no matter what was done to prevent this from happening. Therefore, the idea had been more or less written off. The enemy for all transplant patients—whether they need a new heart, kidney, or hand—is their own immune system. In the case of hand transplants, getting the wrist to accept a foreign hand as its natural extension is the real problem.

The primary job of the immune system is to protect us from invading microorganisms, including bacteria, viruses, foreign proteins, and protozoal parasites—anything that does not belong in the body and thus might potentially cause damage. Primed to destroy all invaders, an ever-vigilant patrol—led by the hound-dog of the immune system, the T lymphocyte or T cell, a white blood cell—is able to distinguish between the molecules produced by these invaders and the molecules produced by our own tissues. The immune system can distinguish between "self" and "non-self."

Sometimes, as is the case with people with AIDS, a crucial part of the immune system has been suppressed and no longer performs its job effectively. Thus, even normally mild infections can take hold easily and pose a much greater danger than normal. Generally, however, the immune system is remarkably effective at sniffing out alien substances. A recognition of foreign material of whatever kind prompts the T cells to send in the hunters and destroy the invaders. Some kinds of lymphocytes secrete proteins, called antibodies, which react with and destroy the pathogen. Other lymphocytes can directly destroy foreign cells or parasites. Still others send a signal to different kinds of white cells and get them to do the dirty work.

Our physiological sense of self is determined by the specific pattern of molecules that stud the surface of each and every cell of our body. What happens when the immune system comes across an entire transplanted foreign organ or body part?

In the case of "transplanted" red blood cells—given, for example, during blood transfusion—that specific pattern is determined by a series of molecules called the red blood cell antigens. There are three main blood antigen types—A, B, and Rh—that, together, determine one's blood type. However, our tissues also have types that are a bit more complicated. Like blood type, tissue type is determined by a series of cell surface molecules. These are referred to as the major histocompatibility complex (MHC) antigens. Unlike blood type, however, MHC tissue type is extremely complex. Because there are far more antigens that vary from individual to individual, it is almost impossible to match the donor type to the recipient type very precisely.

There are two classes of MHC antigens. Class I has some 200 distinct factors. From that each of us selects and displays six factors on our cells; that means there are trillions of unique combinations. Class II is even more varied, with 230 or more variations and eight on display.

So it comes as no surprise that the MHC marking your cells looks entirely different from mine; they are even likely to look different from your mother's or father's. The only perfect match would be between identical twins. But for the rest of us finding a perfect MHC match—one that would make transplanted cells appear to be your own—is virtually impossible.

When the T cells realize there is an MHC mismatch, they initiate a swift immunological response to this sudden appearance of a large amount of foreign tissue. The T cells sniff out a transplanted organ almost immediately, label it as alien, and target that organ for destruction in a process known as rejection.

Rejection is actually two very different processes—acute and chronic. Acute rejection, which takes place within a few days of the transplant, is driven by T cells, and is a fast and savage attempt to expel a new and foreign organ from the body. The T cells set off an inflammatory response, which, in turn, prompts other white cells to swarm to the site and begin destroying the invader. This would have been the kind of rejection experienced by Cosmas's and Damian's patient with the Moor's leg. A limb that is affected by acute rejection would quickly turn black and begin to rot.

That is why transplant patients take daily doses of immunosuppressive drugs, such as cyclosporin. The best of these drugs work much in the same way as the AIDS virus, HIV. They target the T cells and interfere with their ability to transfer signals to the foot soldiers of the immune system. With their lines of communication cut, the T cells stop producing more generations of T cells that might recognize the graft as foreign.

Although putting brakes on the immune system does allow a majority of well-matched transplanted organs to survive, it is a delicate balance, and, depending on the organ, there is a constant threat of acute rejection.

But the real danger in transplantation medicine is the low-level, long-term damage inflicted on grafted organs by what is called chronic rejection. Chronic rejection is a much slower and more subtle attack. It is particularly insidious, in that the worst of the damage is visited upon the blood vessels that feed the organ, causing the buildup of plaquelike scar tissues that narrow those vessels by small increments.

Chronic rejection is mostly an antibody problem; even though the T cells may be persuaded to leave a newly transplanted organ alone, the body still recognizes that something alien has been taken in. In response, it will make antibodies against that alien invader and those antibodies, in turn, begin to tag the organ for destruction, cell by cell. Because immunosuppressive drugs cannot control antibody production, there is less we can do to stop the damage.

All this leads to one inescapable fact—chronic rejection is the reason half of all organ transplants fail within five years.

In the case of hand transplants, surgeons were concerned that the grafted appendage wouldn't even last the first few days. The reason? Conventional wisdom had it that non-self skin grafts stimulate a very strong immune response, and that wisdom certainly seemed to have weight behind it. After all, the scientists said, look at how quickly, how violently, the first hand transplant monkeys had rejected their new appendages. This is why, until the mid-1990s, everybody thought that hand transplantation would fail.

In 1995, Warren Breidenbach, of the Klienert, Kutz and Associates Hand Care Center in Louisville, Kentucky, was determined to try again. Breidenbach suggested simply employing stronger and better drugs, and massively increasing the dose. However, this time, they decided to test their methods on a pig instead of a monkey.

And when Breidenbach and his colleagues began working out methods to deliver huge doses of immunosuppressants into a pig, they found that the increased doses were unnecessary. Pigs given the "normal" amount of the drugs—the amount, relatively speaking, that would be used for a kidney or liver transplant—didn't immediately reject a transplanted limb or hoof. Therefore, when it comes to limb transplants, are we humans more like a pig than a monkey? Breidenbach discovered our response to immunosuppressants is much more porcine than expected. He showed that humans would not necessarily suffer extremely acute rejection, and that one of the last immediate barriers to hand transplantation apparently had been overcome.

After that, it was just a matter of time until the first successful human hand transplant took place. The winner of the race was a French team in Lyon. In September 1998, they transplanted the right hand of a dead 41-year-old man onto the arm of a 48-year-old businessman from New Zealand.

Dr. Breidenbach came in a close second. In January 1999, he grafted a new hand onto the arm of 23-year-old Matthew Scott. The operation, so far, has been a success. Within a few days, there was some slight movement in his fingers; in less than a year, Scott was able to use the hand to open doors and tie his shoe laces.

Then, just after the turn of the millennium, the French team took another step forward, transplanting two hands—and their forearms—onto the limbs of a 33-year-old Frenchman. The patient is doing well, and is starting to get some feeling and movement in his fingertips.

James, who lost his fingers when he was treated with vascular drugs, and Dr. Breidenbach are waiting patiently. When James gets the call that his hands are waiting for him, he will fly on a chartered plane from Denver to the Jewish Hospital in Louisville, where a team led by Breidenbach will perform the 15- to 20-hour-long operation.

No sane person could deny the sheer delight of somebody like James having a new pair of hands. Clearly such treatment would produce a vast improvement in James's quality of life. But there are difficult and controversial ethical questions raised by this kind of transplant. Apart from the

Clint Hallam, one of the first hand transplant patients, enjoying the company of his family.

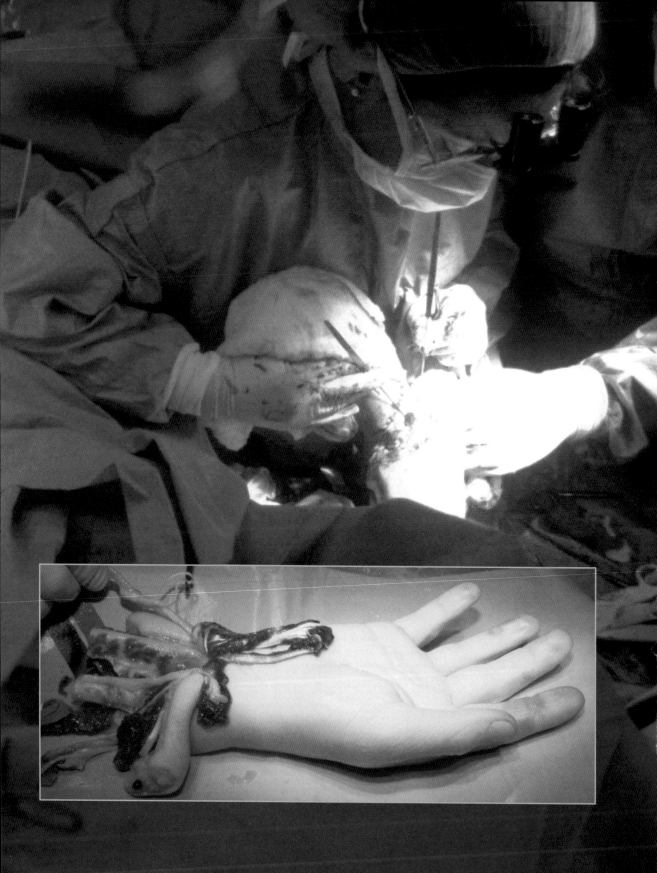

issues that concern the source of the donated limb, James is running a very substantial risk, which many people feel is not justified. Yet again, there is the uncomfortable feeling that the doctors concerned enjoy the race, enjoy the achievements, and enjoy the fame that their surgical prowess may bring, perhaps at someone else's cost.

In addition, of course, James could end up with a "dead" hand, devoid of proper sensation—a hand that, for example, could get burned or trapped in doors because he may not know where the hand is without looking at it. Even if James gains reasonable sensation and mobility, which is admittedly more likely now with improved surgical techniques but by no means guaranteed, he will have to contend with a substantial chance of chronic rejection. He may also anticipate an earlier-than-normal death because, over a period of five to ten years, the immuno-suppressive drugs needed to prevent acute rejection will damage his immune system to the extent that his life expectancy will be drastically reduced by an estimated 20 years.

As mentioned, the dangers of using these drugs are similar to immune deficiency diseases such as AIDS; the body is much less able to fight off infections. There is also a substantially increased risk of cancer. Without a vigilant immune system in place, a transplant recipient's body is thought to be around three times more likely to let a stray cancer cell slip by unnoticed. Some researchers think it is possible that immunosuppressants, such as cyclosporin, may promote the growth of tumors. Indeed, if a person lives long enough on immunosuppressive drugs, his or her chance of developing, for example, a skin cancer can be around 90 percent. These drugs also make diseases such as diabetes more likely.

This is the heart of the ethical dilemma. Breidenbach concedes that hand transplantation is an "elective procedure"—in other words, it is not done as a matter of life and death. A heart transplant patient would die without a new heart; James can live without his hands. However, he will be taking the same kind of immunosuppressive drugs taken by heart transplant patients, and many people believe that pumping an otherwise healthy person full of chemicals in order for them to accept a nonessential organ that has a strong possibility of not working properly is the height of folly. Can both patient and doctor justify the increased risk of infection, and of cancer, and hence a significantly reduced life expectancy?

For most people, the decision to undergo a hand transplant would be harder than the decision to opt for a heart transplant. But Breidenbach

OPPOSITE *Hooking up a hand is a complex business. Not only are there numerous tiny vessels to rejoin, but the tendons that move each finger and the nerves that control the muscles all have to be reconstituted, a highly technical procedure.*

points out that kidney transplants, which are now routine, involve a similar decision. A patient can live without a kidney, but this person will have to spend the rest of his or her life on dialysis, which severely affects the quality of life. The question is whether a similar loss affects those with no hands, and Breidenbach is convinced that this is the case. He says it is unfair to dismiss the psychological and emotional effects of losing one's hands; we should not think of a hand transplant as the high-tech equivalent of a nose job.

"Think of the loss of both hands," he urges, "not only about what you can't do, but how you look and how you interact with people. When you reach out to shake someone's hand. When you go to hold your children. When you go to embrace someone and kiss them. The consequences can be devastating. There are people who do quite well even if their hands are missing. They get prostheses, get on with their life, and do okay. There are others who have tremendous problems."

James has had problems. His prosthetic hand, which is a simple hook, is uncomfortable to wear and sometimes painful. It provides him with little grip strength; he cannot grasp a knife firmly enough to cut through a steak. He does not have enough dexterity to button his own buttons. In order to crack open an egg, he has to smash it, then lift out the pieces of shell. James puts his prostheses on only to accomplish quick, discrete, simple tasks. He rarely, if ever, sits around with his prostheses on for hours at a time. Some days, he says, he never even puts them on at all.

He knows that a hand transplant is not going to put things back to the way they were. The doctors have cautioned him not to expect full function in the transplanted hand, nor a return of his fine motor skills.

"Still," says this engineer by trade, "four fingers and a thumb, even on a limited basis, is one hundred times better than any of the prosthetics they've come out with today."

James makes a simple request of those critics who say such procedures should not even be available, those who say the risks outweigh the benefits: "Put yourselves in my place."

"They have to realize that standing there with two hands saying, 'I can't believe someone would take these drugs to have hands,' well, it's very easy to say that. But when you don't have hands, the drugs seem like a minor footnote."

Transplantation fascinates because it blurs boundaries. If a hand that belonged to another person sits at the end of my arm, will it ever be truly part of me? If I carry your heart in my chest, do I also carry your essence, your spiritual core?

Barney Clark with his Jarvik–7 beating away in his chest showed no signs of turning into an android. Body parts are, after all, only body parts made of flesh, bone, and tissue. But organs do share an identity and a biological fingerprint with the donor. Rejection taught physicians from Cosmas and Damian onward that body parts cannot be interchanged between people at will. Every heart or kidney retains a physical connection with its previous owner, and this deep-rooted identity is often at odds with the recipient. This may make psychological sense. I have heard it repeatedly said that many people find it quite difficult to shake hands with someone who has had a hand transplanted—the notion that they are in contact with a dead person overwhelms any other feeling.

Physiologically, the seat of our identity is not in the heart, nor even in the brain, but literally in the very marrow of the bones. The bone marrow's stem cells ultimately differentiate between self and non-self, covering the immune system's T cells with receptors that recognize self-created tissue, bone, and blood as friendly and anything else as a potential danger.

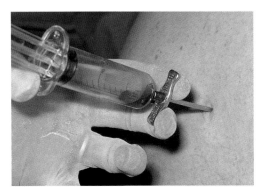

When an organ from a stranger is introduced, those same patrols will tag it as foreign and set out to destroy it. If we wipe out that immune system by replacing all of the body's bone marrow, and replace this with a foreign version, the bone marrow graft is likely to destroy its host, the very body that it needs to survive. That is how "single-minded" the immune system can be. Transplantation, therefore, presents a unique challenge; we need to expand the "mind" of the immune system, persuade it to accept foreign tissue, while still maintaining its necessary degree of vigilance to fight off true invaders. One potentially very successful way to do this is by creating a chimera.

The fire-breathing Chimera of Greek mythology had the head of a lion, the midsection of a goat, and the hind parts of a dragon. John

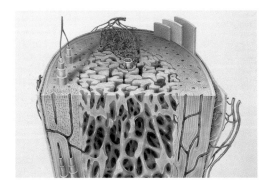

The marrow (above) is highly active, and it makes new red and white blood cells. It is generally harvested for transplantation by inserting a needle into the bones of the pelvis (top), which is a particularly active factory in most adults.

Otting, Sr., has none of these, but the 64-year-old retired radio station owner might become a chimera of sorts. He is planning to undergo an infusion of stem cells into his bone marrow, cells donated to him by his son, John Otting, Jr. His son will also donate a kidney. The operation is primarily designed to revive John Sr.'s failing health and keep him off dialysis, but if the bone marrow transplant takes, John Otting, Sr., could become a human chimera, and the problem of rejection overcome.

John Sr. has been a diabetic for most of his life and has suffered the kidney complications that are not uncommon in people fighting this chronic condition. By November 1998, he was severely anemic and constantly exhausted. Without a transplant, dialysis would have been unavoidable, and he could not bear the thought of being tethered to a machine for hours upon hours, week after week, for the rest of his life.

> "I'm two persons in one, … my son is me—he has my DNA—but I, in turn, am him. I am the father and the son. I have his bone marrow. I have his kidney. He will be with me for eternity."

Rather than opt to get on a waiting list for a cadaveric kidney, John and his family decided to consider a living transplant. Against all odds, John Sr.'s wife of 42 years was a reasonable tissue match. Unfortunately, she developed kidney stones soon after testing, which knocked her out of the running. Both of the sons were willing to consider a transplant; it was John Otting, Jr., who got tested first, and he was found to be a match.

Then, the Ottings's surgeon, Frederick Bentley, mentioned the possibility of creating a familial chimera.

Bone marrow transplantation is different from almost any other type of tissue or organ graft. Bone marrow is the only body tissue that, once it is transplanted and takes, is permanently accepted without requiring immunosuppressive drugs. Crucially, if the bone marrow takes, then a heart or a liver or a lung from that same donor will also be accepted with little danger of rejection.

In the world of transplantation, this is called tolerance. A person who has received another person's hematopoietic stem cells—the cells that produce red cells, platelets, and the majority of the white cells—will begin to pump out T cells that will "tolerate" tissue from the same donor. If we then transplant that donor's kidney—or pancreas, lung, or heart—into the same recipient, it appears to be accepted as friend rather than foe. This merging of two bone marrow identities results in an immunological chimera. We can create a person with two sets of stem cells and the ability to create two complete immune systems that have the ability to recognize two distinct biological identities.

In the case of John Otting, Sr., chimerism meant giving him a bone marrow graft from his son, John Jr., along with his kidney transplant. Probably 90 percent of the older Otting's bone marrow would still be his own. However, if one were then to test his blood, his particular tissue fingerprint would appear to be a mixture of both himself and his son. He is a chimera.

"I'm two persons in one," John Otting, Sr., smiles. "I mean, my son is me—he has my DNA—but I, in turn, am him. I am the father and the son. I have his bone marrow. I have his kidney. He will be with me for eternity."

Dr. Bentley claims that a chimeric immune system does not require any immunosuppression, with its myriad side effects and risks. The chances of long-term or chronic rejection of the donor organ is reduced, if not removed entirely. There are many potential benefits to be had. Fewer people should lose kidneys after nine or ten years and have to return to the transplant list; more organs would be available to be transplanted; the cost would go down.

The risks, on the other hand, are relatively minimal. Just a few hours before the start of his kidney surgery, Otting, Sr,. was given a relatively low dose of radiation to kill off just enough of his bone marrow cells to make room for his son's. Not many donor stem cells are needed to get the message across; if just one percent of a person's immune system derives from the donor, then it should shift the immunological sense of self sufficiently to allow a kidney transplant from that same donor to thrive unmolested.

For Otting, Jr., who was already undertaking the risk of kidney surgery, the adding of a bone marrow transplant to the mix would simply result in aching pelvic bones in addition to a painful abdomen. In a family as close-knit as the Ottings's—the boys worked with their father at the family-owned radio stations—the idea of this peculiar melding of identities was more fascinating than it was off-putting. Otting, Jr., says, "We've got the same name. We've always been pretty much one of a kind." Now, if the marrow transplant takes, they really will be one of a kind.

Chimerism began as an attempt to address a terrible condition called graft-versus-host disease (GVHD). In a cruel twist of fate, bone marrow recipients, whose own immune systems have been destroyed by drugs and radiation, are attacked from within by their donor's healthy cells. This kind of reverse rejection occurs in about 50 percent of all patients who receive marrow from a related, matched donor, and in even higher percentages of patients whose donors are not related by blood. Their own bone marrow starts to reject them.

Acute GVHD often begins with a skin rash, which is seemingly benign to the untrained eye but actually a sign that the new immune system is intent on destroying the recipient's body. GVHD does not stop at the skin; the condition can attack the gut, destroying the lining and causing devastating bouts of diarrhea. It can debilitate the liver, destroying its ability to process toxins. It can settle into the lungs, causing pneumonia or bronchitis. In many cases, it will also kill.

Suzanne Ildstad, Director of the Institute for Cellular Therapeutics, University of Louisville, Glenolden, Pennsylvania, and a professor of surgery, is one of those doctors who have set their sights on banishing GVHD entirely.

With a near-perfect bone marrow match from a sibling, the recipient has a 40-percent chance of developing GVHD. If this happens, there is about a 20-percent chance of dying from the condition. With a five out of six match, there is a 50-percent chance of developing GVHD, and subsequently a 50-percent chance of dying. When four of six characteristics match, there is a 100-percent chance of GVHD and, therefore, a 100-percent mortality rate.

So, even if it is possible to get a perfect match from a sibling, the risk is still unacceptably high. To give the host a fighting chance against the bone marrow graft, scientists began to explore the possibility of removing the mature, fully armed T cells from the donated marrow—a process called conditioning. Without these most reactive of T cells, the grafted marrow is unable to launch an effective attack on the host. In a sense, it has been disarmed, at least temporarily. The problem with this approach, however, is that harmless marrow turned out to be useless marrow. They reasoned that the same cell that caused GVHD was the one required for the bone marrow to take.

It was a reasonable conclusion, but, as Dr. Ildstad soon proved, an incorrect one. The T cells that cause graft-versus-host-disease reaction, it turns out, are not the same cells that allow engraftment. In fact, the cells that allow engraftment, cells that Ildstad discovered and called facilitator cells, are not T cells at all. They are simply a sort of stem cell caretaker, creating the best possible environment in which the stem cells can settle down. They are the reason that stem cells do or do not "take" when grafted. They were never noticed before, says Ildstad, because they too are destroyed by the chemicals used in marrow conditioning, along with the T cells that cause GVHD.

For a bone marrow graft to take well yet avoid GVHD, the right balance between these three cells must be found. In other words, the best bone marrow for transplantation is loaded with stem and facilitator cells, and has just enough T cells to get the best possible graft without that graft attacking its new host.

Preparing donated bone marrow for one of Ildstad's tolerance trials is a painstaking process. Donated marrow is couriered at high speed to the laboratory and processed by a team which is on call 24 hours a day, seven days a week. These specialists then spend the next 18 hours winnowing out the undesirables—the mature, fully equipped T cells that could cause GVHD—and leaving everything else behind. The treated marrow is then carried back to the patient.

The first tests of Dr. Ildstad's technique involved 70 leukemia patients for whom a perfect or near-perfect marrow match could not be found. Leukemia patients have a cancer of the bone marrow; in many cases, unless the marrow is destroyed, these patients will die from the leukemia. One desperate treatment that is normally given involves total body irradiation to destroy all of a patient's own marrow.

To undergo this treatment, several hours of intense irradiation is required, and that is like sitting near a nuclear explosion. Thereafter, the patient needs a transfusion of donated marrow; otherwise, he or she will die within days of this dramatic treatment. These patients were those whose marrow had only three out of six characteristics match the donated marrow. In the past, this would have meant certain GVHD, and probably death. Instead, many of these patients are going home after just two or three weeks.

After that resounding success, humbly known in the medical world as "proof of concept," Dr. Ildstad has been given the go-ahead for a number of larger-scale trials, several of which involve patients without leukemia. These patients have nonlethal diseases, including blood disorders such as sickle cell anemia or thalassemia. Dr. Ildstad's team is also working on transplant tolerance trials, in which it is attempting to create chimeras and trick the body into considering a donated organ as its own in cases where the bone marrow and tissue match is less than perfect.

Disappointingly, in the case of John Otting, Sr., the bone marrow graft has not taken. As a result, he will have to take immunosuppressive drugs for the rest of his life. But even though the marrow transplant failed, Otting, Sr., still has his son's kidney, and hope for the future. His words remind us of why we are devoting huge amounts of money and some of the best minds in medicine to transplant research.

"It's like being reborn. My son gave me life, and a real life. Because of him, whatever years I have left, they will be quality. I will travel, I will enjoy my life, and I will not be a kept person, bound by dialysis. To have somebody who loves you enough to do that is astounding."

Chimerism may be a potential solution to the problems of rejection and the dangers of long-term immunosuppression, but it does not solve a pressing problem—the organ shortage. We can create all the chimeras we want, but if there are no available organs, people are still going to languish and die on waiting lists.

Therefore, to increase the pool of available organs—a very shallow pool, at that—researchers have turned to animals. Xenotransplantation, as the practice of harvesting body parts from one species and transplanting them into another is called, is potentially the best and most audacious answer to the lack of human organs.

If chimerism plays around with our human identity, xenotransplantation would seem to completely destroy it. Is a human being still a human being if he or she has the heart of a pig? And what if a person has the immune system of a baboon?

In 1995, Jeff Getty was not overly concerned about his self-identity. Getty, an AIDS activist from Oakland, California, was a very sick man who was waging a debilitating battle against the human immunodeficiency virus. His T cell count was pitifully low, and his body had declared open season for any opportunistic infections that wandered by. The triple-drug protease cocktails that had raised so many hopes among the HIV community had failed him. Getty's options were dwindling.

Then Dr. Ildstad offered him one more chance. It was a radical concept, to be sure, but Getty has always thrived on the radical. Ildstad proposed a bone marrow transplant, in which the donor marrow would be sucked out of the bones of a baboon. Because baboons are one of the only primate species that are resistant to HIV, the hope was that by injecting Getty with the essence of a baboon immune system, he too would produce immune cells capable of quashing the virus.

Getty was to be the very first attempt at melding man and baboon, at least immunologically, in an effort to destroy the AIDS virus. The renowned transplant surgeon, Thomas Starzl, had previously transplanted baboon livers into two AIDS patients suffering from hepatitis.

Unfortunately, both had died soon after their procedures in 1992 from infections associated with the extreme immunosuppression required to get a human body to tolerate baboon cells.

This time, however, things would be different. Armed with the knowledge of her facilitating T cells, Ildstad attempted to induce chimerism without requiring serious immunosuppression in an already oversuppressed system.

Of course, immunosuppression could have been the least of Getty's problems. There was always the chance that the baboon donor might be carrying an unknown virus in his or her blood, one that could infect Getty and kill him. But there is also another very serious concern—a real worry that an animal virus infecting a human, might undergo change, and that such a change might make the virus widely infectious enough to spread to the world at large. Indeed, the danger of transferring animal diseases—so-called xenozoonoses—into the human population has been one of the key worries holding back the field of xenotransplantation today.

Ildstad thought it was important that Getty should know what he might be letting himself in for. If he accepted the baboon cells into his body, his social contacts would be severely limited for the foreseeable future; his blood would be tested in an obsessive schedule to make sure no unidentifiable infections were gaining a foothold. He would, in short, become one of the most-watched, most-studied people alive.

There are vocal opponents of xenotransplantation, who say that the risks of xenozoonoses infecting the human population are far higher than the rewards of the actual transplant. Ironically, they cite AIDS as a prime example; HIV probably evolved from a monkey immunodeficiency virus that, in unknown conditions, jumped from primates to humans.

Although Dr. Ildstad was prepared to take all possible precautions, she does not believe the risks outweighed the rewards. Getty listened to Dr. Ildstad's ideas and her cautions, considered them, and accepted. He had nothing to lose. And so, on December 14, 1995, at San Francisco General Hospital, doctors injected baboon bone marrow into Getty's bloodstream.

The first attempt at xenotransplantation was back in 1906. A French surgeon, Mathieu Jaboulay, transplanted a goat's liver into one woman and a pig's kidney into another woman. Neither woman survived the operation. No one dared try again until 1963 when there were six attempts at kidney transplants between chimpanzees and humans. The longest survivor lived for nine months. This was a particularly striking result, given that efficient immunosuppressive agents had yet to be developed.

Perhaps the most notorious of the attempts at xenotransplantation was the one made on Baby Fae. In 1982, this infant received a baboon heart to replace her own failing organ. Her physicians were hopeful that a new antirejection drug, cyclosporine, would give her a good chance of survival. Nevertheless, she died less than three weeks later, after rejecting the heart, and the highly publicized case cast a pall over the field for quite some time.

In Jeff Getty's case, Ildstad watched on tenterhooks as the baboon cells gained a tenuous foothold in his bone marrow. Then, for reasons still unknown, they were nowhere to be found just a few weeks later. Getty was no chimera; his immune system was his own again.

Officially—scientifically—the Getty experiment was a failure, and yet Jeff Getty thrived after his transplant. He gained 20 pounds. He stopped getting sick. His viral load—the number of HIV particles detectable in a milliliter of blood—dropped to zero for almost a year.

Dr. Ildstad cannot explain why. When the virus once again made itself at home in Getty's body, he continued to fight off the infections successfully. Had the baboon engraftment, even though it lasted for two weeks, played a role, first in eliminating the HIV and then in Getty's ability to deal with the virus once it returned? There is, as yet, no answer.

But there is one thing Dr. Ildstad is sure of. Getty showed no signs of a xenozoonotic infection.

Despite Dr. Ildstad's setback, xenotransplantation is going from strength to strength. Those who champion the cause say that this is the solution to the lack of human organ donors. The critics, however, say that we simply do not know enough about the transfer of animal viruses to the human population. For example, they point to an incident in Malaysia in 1999, when over a 100 people died after contracting an unknown virus that mosquitoes were thought to have carried from pigs.

Even if we take the greatest care in breeding disease-free animals, we cannot avoid animal viruses completely. For example, all pigs carry in their DNA a virus that is an integral part of their genetic makeup—a so-called retrovirus, which would always be transferred to humans if pig organs are used in transplants. These retroviruses are the subject of close and nervous study.

There is an undeniable risk, but there has been a great deal of research showing that, among patients who have already had pig tissue implanted in them for years, there is no sign of any

OPPOSITE *Baby Fae, who made headlines in October 1984, when she received the heart of a baboon and survived for less than three weeks. To insert the heart her chest was split right down the middle; to prevent infection she was kept in an isolated incubator.*

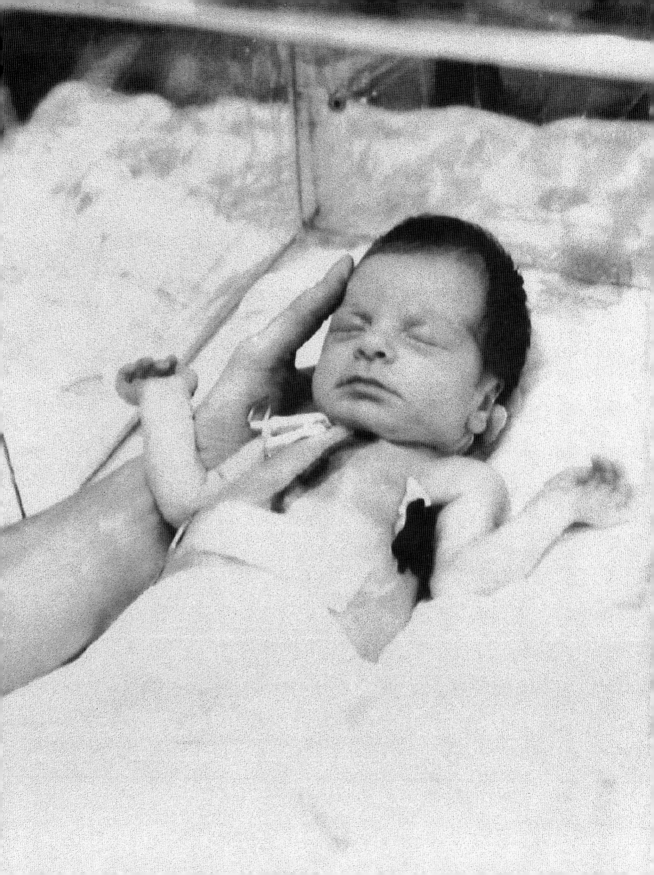

dangerous infection. Where the retrovirus is present in humans, it is because pig cells remain in their bloodstream; it does not seem to have affected their health.

The main barrier to xenotransplantation is, once again, rejection and in particular a phenomenon called hyperacute rejection, which occurs because the two species are not closely related. This is a brutally fast reaction that causes the blood in the implanted organ to coagulate so that the organ is immediately starved of oxygen and dies within minutes.

Key research in the field has been done at my own hospital, The Hammersmith. There Professor Robert Lechler heads a unit that has been making great progress. He has linked up with Imutran, a UK company based in Cambridge. They have developed an ingenious but controversial solution—they have bred genetically modified pigs.

They inject the pig embryo with genetic material that contains the code for certain human proteins. The proteins find their way to the surface of the pig organs, which makes the organs appear to the human immune system as *human* organs. The T cells cannot tell the liver or heart apart from a human organ; therefore, hyperacute rejection is prevented.

Immunosuppressant drugs are still needed to prevent normal acute rejection, just as they are needed in human-to-human transplants, so the dangers of taking these drugs over a lifetime have not been avoided; there will still be increased risk of cancers and infections. But Imutran's work means that pigs can be bred especially to produce hearts for transplanting into humans; if the technology works as it promises to work, there could be more than enough hearts and other organs to go around. Professor Lechler tells me that he believes the problem of hyperacute rejection has been solved.

My own research may come in handy at this point. One of the difficulties with making transgenic pigs—pigs carrying foreign genes—is that the technique is very expensive. Pig embryos are hard to come by and, after manipulation, very few implant. The technique for modifying the sperm that I describe in Chapter 6 may be very valuable here. In theory, this could mean that pigs suitable for human transplants could be bred by natural mating once we had injected the right genes in their testes.

Another alternative might be cloning. I am less optimistic about cloning than many colleagues. Cloning is very labor-intensive and very inefficient. Many animals born after cloning have unexplained abnormalities and may be unsuitable for human transplants. But time will tell; I believe the future of this branch of medicine is looking distinctly rosy.

The mass farming of pig organs is probably still many years away, but the ethical debate is in full swing. Research into xenotransplants is, as previously mentioned, running the risk of animal viruses that could, in theory, unleash disaster upon the human inhabitants of the world. Then there are the moral implications of breeding genetically modified animals to provide organs for humans, which many critics find troubling. However, I feel that if we are prepared to eat meat and wear shoes made of leather, neither one being essential for comfortable human existence, the use of animals to save human lives is completely justifiable.

The residents of Mount Sinai Hospital would be delighted to have any healthy heart, whether human, pig, or plastic.

Peter Liuzzo got his new (human) heart on November 10, 1999, nearly eight months after he first checked into Mount Sinai for tests. He remembers his first day on the ward.

"I could never join the group. I actually told them, 'You guys are sick. I don't wanna have anything to do with you. Because in a few days I'll be going home.' Well . . . a few days was eight months ago."

It took nearly six hours and six highly trained specialists—two surgeons, an anesthetist, a circulation nurse, a cardiologist, and a bypass technician—to give Liuzzo a new life. Nevertheless, only two weeks later he was home, just in time to celebrate Thanksgiving with his family. He certainly had a lot to be thankful for. In almost a matter of days Peter was barely recognizable. He had been transformed from a drawn, gray, frail old man into full-faced, ruddy 62-year-old fellow.

Dolly the sheep, almost as interested in the gentlemen of the press as they are in her.

He says, "If somebody was passing on and wanted to leave a gift for humanity, this is the gift. Whether it gives a man five years, three years, ten years, it's something we didn't have before. We were condemned men, all of us."

Today, in Mount Sinai Hospital, there is a new generation of faces looking out on Central Park. Just like Peter Liuzzo and John Weissman, they support one another, cheer for one another, pray for one another. And complain about the food.

3 Self-repair

Penny Roberts says her 351st sky dive felt just like any other. But it was the first time she had been up in a plane for a year. And this time, unlike all the others, she had to jump with a partner—someone who would walk her to the airplane's door, jump out of the plane with her, and reach up and pull her parachute cord.

Penny, a 37-year-old nurse from Steeton, a small town on the moors of West Yorkshire, England, is paralyzed from the chest down, the result of a sky dive that went terribly wrong. But she refused to let that ground her. The first jump after the accident was a memorable moment:

"I remember seeing my wheelchair on the ground getting smaller and smaller, and that was brilliant. When we got up to 15,000 feet, I remember being in the door just before we exited from the aircraft, and I looked down and saw my feet, and saw the clouds below my feet. And I just thought, 'I'm back.'"

Penny's first-ever jump was to raise money for the British Heart Foundation. It was a gutsy move—she had never been in a plane before. In fact, she jumped out of a plane 11 times before she ever even landed in one. But, after that first time, within seconds of leaving the aircraft, she was addicted: "I just wanted to go straight back up and do it again."

So she did. Penny's accident happened some five years later on her 350th jump. It was in the spring of 1995, and Penny and the team had flown to Florida to train for some domestic sky-diving competitions coming up that summer.

The training was going well. On the final jump, they squeezed into the Cessna, taxied down the runway and soared up and away into the blue skies. At 15,000 feet they leaped out of the airplane, maneuvered into position, and practiced their elegant aerial spins as they fell through the atmosphere at 120 miles per hour. A cameraman filmed the team from above so they could analyze their performance on the ground. Then they reached 3,000 feet and moved apart to open their parachutes. Penny pulled the rip cord. The pack opened, the parachute flapped and fluttered, but it did not open.

Penny tried to jettison the faulty main parachute to open the reserve, but it had snagged on her shoulder harness. The two canopies became hopelessly entangled. The cameraman put himself into a fast spin to try to catch her but he could not make up the ground. Curiously, Penny did not feel any fear; she spent the time thinking about what she could do to improve the situation. Unfortunately, the answer was absolutely nothing.

The videotape records everything that happened, right up to the moment when the cameraman landed and ran up to her on the tarmac. It

Penny Roberts jumping in tandem with her instructor from 15,000 feet above sea level.

reveals the crumpled white canopy doing very little to slow Penny's fall, and when her body hit the ground she was traveling at more than 60 miles per hour.

Penny suffered a fractured skull, a small brain hemorrhage, two broken ribs, two collapsed lungs, a broken pelvis, and a broken right leg. Six of the eight vertebrae in her neck were shattered, as well as two more in her back. Her spinal cord was severed between the fifth and sixth vertebrae, leaving her paralyzed from the chest down. She was extraordinarily lucky to be alive, but her body was damaged beyond repair.

Penny did remarkably well, all things considered. She endured dozens of operations and dangerous infections. Her X rays now reveal a mass of metal that holds her broken bones in place. Five years later the bones have healed, her lungs are healthy, and her brain suffered no lasting damage. However, the severed spine and the paralysis seem irrevocable.

This kind of injury is devastating. It is sudden, one immediately feels helpless and dependent on others, and there is very little or nothing that surgery and medication can achieve. To get back on her feet, Penny would need to coax the severed nerves in her spinal cord to grow again, to cross the gap in her broken back and then send the right signals to the right places. She would need to regenerate this delicate and highly complex bundle of nerve cells. Unfortunately, evolution has simply not given our bodies the tools to carry out these tasks.

A salamander is a most extraordinary animal. When it loses a leg in an accident, or has it torn off by a predator, we would expect the wound to simply heal. But, in a matter of days the stump begins to grow. Bone and muscle gradually extend outward; nerve cells and blood vessels grow, weaving their way into this new structure. Just a few weeks later, the toes can be seen taking shape and the scaly skin hardens around them. The salamander can grow a leg that is a perfect replica. The new leg has all the power and function of the old one and the salamander is as good as new.

The pondworm, *Lumbriculus variegatus*, can go one better. If it is chopped in half, each half regrows into a whole, complete worm. This concept, when we stop and think about it, is extremely bizarre. One animal becomes two, just like that. Which half gets the brain? Which gets the stomach? Which the ovaries? Is it still one animal, which just happens to be made up of two separate parts?

A worm sliced in two is not one dead worm but two live ones. Each half of the worm, now a kind of identical twin, spontaneously regrows the missing half of its nervous system and bodily structure. The worm is, admittedly, a relatively simple creature with no real brain to speak of. Nevertheless, it has highly specialized tissues, reproductive and other organs, nerves, a mouth, a bowel, and an anus, so this is still the most tremendous feat.

In some respects we humans, too, are pretty good at regeneration. We are constantly growing new cells to replace those that are destroyed. We cast off skin cells continually, shedding and replacing some 500 million each day; over a lifetime, each of us loses about 40 pounds' worth of the stuff, quite a lot of which ends up as house dust, and in our bedclothes. White blood cells live lives that seem independent of the body they inhabit. They swim around in the bloodstream quite vigorously and then die after only a few weeks.

OPPOSITE The salamander is a remarkable beast with an extraordinary ability to regenerate its limbs.

Red blood cells can ferry oxygen for about six months before needing to be retired, but we still replace about eight million of them each minute. Our hair lives longer—each strand hangs around for two to four years. And our bone cells can live for up to 30 years, although the average is ten. A few nerve cells—for example, those that stretch all the way from our legs to our brains—last our entire lifetime. But elsewhere, the rule is constant renewal.

This all goes to show our remarkable human talent for keeping our bodies functional over an extraordinarily long life span. But for sheer biological showmanship, we will never match the prowess of the salamander and other reptiles. Hack the tail off a lizard, and it will be whole again in no time. We cannot replace a severed arm—nor, for that matter, a severed hand or finger, or even the tip of a finger.

> … Hack the tail off a lizard, and it will be whole again in no time. We cannot replace a severed arm … a severed hand or finger, or even the tip of a finger.

The further down the evolutionary hierarchy we go toward the reptiles, amphibians, and worms, the more widespread this regenerative talent appears to be. And peering down at the taxonomic lower classes, we are, in some ways, gazing back into our own evolutionary past. Our direct ancestors, after all, include something akin to worms, as well as basic vertebrates that may have been something like the salamander. It is certainly possible that these ancestors also had the capacity for regeneration.

It seems somewhat paradoxical, then, that over evolutionary time, we have lost the ability to replace body parts. Evolution always favors any change that gives an organism a survival advantage. Why would natural selection favor *losing* a limb over *replacing* that limb?

The answer is that no one really knows the answer. There are various theories. One such theory points out that regeneration is fabulously expensive, from a metabolic point of view. For large mammals, primates, and early humans, perhaps regeneration simply did not pay off. Just think, compared to a salamander, how much more energy it would take to rebuild a human leg and all of its tissues. In addition, think how much time such a process would consume. A salamander takes just a few weeks to grow a leg; we would surely take months to accomplish the same task. Even if we tried to regenerate a leg as a young child, the sheer size of our bodies, and the diversion of precious resources, would certainly weaken us, affecting our general health, as well as simply requiring more food.

Another theory implies that during the process of regeneration we would be much more vulnerable to predators. Initially, the bone and muscle would be weak and delicate. A regrowing leg would not be functional for such a long time that, during the early stages at least, we would be no better off in terms of mobility and our ability to run away from the saber-toothed cats. We would be worse off, in fact, because we would be diverting precious energy, oxygen, and food to the new leg. There is also the question of infection. A regenerating proto-leg, with its fragile new skin, would be more susceptible to attack from invading bacteria and viruses.

From an evolutionary standpoint, then, perhaps the most sensible response to the loss of a limb is to close the wound as quickly as possible. That way, the flow of blood is stanched and the body is protected from microbial invaders. So we patch up our wounds with scar tissue.

Scar tissue is a make-do, quick-fix compromise. It is knotty, quickly knitted, and crude, with none of the elasticity, the subtle biochemical functions, or the look and feel of normal skin or organs. But, by simply slapping a quick bandage of skin over the wound or stump, the early human did not spend all of his energy trying to build a leg that would not have been able to hold his weight when he was faced with a threat. Perhaps this is all that it took to tip the evolutionary scales, to favor the formation of scars rather than the regeneration of limbs.

A serendipitous clue led to another theory, a theory that claims that the powers of regeneration are related in some way to the immune system. A few years ago, a Philadelphia researcher, Dr. Ellen Heber-Katz, bred some mice for an entirely unrelated experiment. As part of normal laboratory procedure she had marked them, for identification, by punching a small hole in their ear. Three weeks later, she examined the mice and was amazed to discover that the gap had been completely filled in with cartilage. Normally the holes might be fringed with scar tissue, but these mice had outdone themselves. They had regrown their ears perfectly, with normal, supple cartilage, which had even a grown a network of blood vessels.

This was not supposed to happen. Amphibians and lizards can perform these feats of regeneration. Birds and mammals cannot.

Other strains of mice that Dr. Heber-Katz tested did not regenerate in this way. But these were no ordinary mice, they were of the MRL strain and specially bred with no immune system. Dr. Heber-Katz thought this must be the key to this unusual ability, so she and her colleagues looked at the pattern of inheritance in MRL mice. They eventually found that healing ability in these animals was linked to particular areas on different chromosomes, presumably the influence of certain genes. Their lack of healing was also in some way linked to the fact that they were immune deficient.

Perhaps, the secret of healing lies in our lack of an adequate immune system. If we are still genetically capable of regeneration, like these trailblazing mice, perhaps our immune system is repressing the genetic networks responsible. These genetic networks may still be latent within our DNA, waiting to be roused from millions of years of slumber. Maybe there has been an evolutionary trade-off. Perhaps we humans need such a complex immune system that we have traded the ability to regenerate for the ability to fight pathogens and cancer cells.

Plausible evidence exists in support of this idea. Regeneration and tumor formation, both of which involve cell division and migration on a large scale, are very much alike in terms of the chemicals they produce and those they require. For the immune system to scout out tumor cells vigilantly and protect us from the dangers of cancer, it would also have to close down most major regenerative projects.

> So we turn to medical intervention … Can we coax our chemical messengers to issue orders to renew and regenerate the body?

So much for evolution. It seems unfair that some upstart amphibian can perform a feat that for us is just a frivolous daydream. However, medical science is not completely daunted by our evolutionary limitations. There is an especially intriguing question. Might it be possible to restart these processes artificially? Is there any way we, as a species, can remember how to perform these biological tricks? Can we flip the genetic switches and power up our capacity to regenerate?

If the mice, with no immune system, can do it . . . Well, there is just one time in our life when we humans do not have an immune system—indeed, the only time that every mammal lacks an immune system—and that is when we are embryos.

Every one of us used to have the ability to generate and regenerate body parts; as embryos, just days or weeks old, we were able to create bones, blood vessels, organs, even arms and legs out of a single fertilized cell. The genetic cascades, which turn these processes on and off, are at their most powerful when we are embryos and the clues guide us back into the womb to our earliest days of life.

At that stage of development, we can also repair lost body parts. If an embryo is damaged at a very early stage, when it is just a clump of cells, if one or two of those cells are torn away, then more cells will grow in their place and the embryo will develop whole and fully formed. If, during the first ten or so days of existence, this clump of cells is split into two, then like the worm each new clump will become a fully functional identical twin. But, somewhere along our journey from egg to infant, we lose the capacity to rebuild our bodies on a large scale.

There are some important exceptions. For example, adult human beings can lose as much as two-thirds of their liver and, in just a few months, it will grow back again, without any loss of form or function. But an amputee will never find a new limb spontaneously growing in place of the old one. It is unlikely that Penny Roberts will regrow her spinal cord.

So we turn to medical intervention. What if we can persuade our own bodies to repair dying organs, or even construct new ones? What if nerves can be regenerated, and brains remodeled? Can we coax our chemical messengers to issue orders to renew and regenerate the body?

There are a great many scientists who think they have discovered the secrets of regeneration, and the embryo is where their inspiration is to be found. They watch carefully as a clump of cells manages to coordinate and construct every complex organ and physiological system in the body. They are starting to believe that if these cells can build, surely they can rebuild; maybe those developmental processes can be reactivated in order to repair a diseased or broken body. If so, that would be a truly Superhuman feat. That would be worthy of any salamander.

Reverend Charles Wilson has a condition called familial hypercholesterolemia. His maternal grandfather had it and passed it on to Charles's mother (and her nine brothers and sisters). She, in turn, passed it on to Charles, who has passed it on to all six of his own children. Inherited hypercholesterolemia is common—indeed, probably the most common genetic disorder in the United States and Britain.

It is a most unwelcome inheritance. Somewhere in the twisting strands of DNA at the center of each of his cells is a muddled gene on the 19th chromosome, which signals his body to build up cholesterol at an appallingly high rate. The fatty substances that result are deposited straight onto his arteries, where they form thick yellow plaques. These deposits are called atheroma. With them, the arteries get narrowed and easily block up. And the blockages are usually most frequent where blood flow is most important—the vessels of the heart and those supplying the brain.

Charles used to be slim and athletic; he wrestled in high school and college, ran five miles a day, and lifted weights. By the time he was 27 years old he had the arteries of an obese, sedentary 70-year-old man. Around that time, he experienced his first severe angina attack, an excruciating pain accompanied by a feeling of suffocation, which was caused by a lack of oxygen to the heart. It is the kind of pain that makes you think you're just about to die.

In the 30 years since, Charles has had an untold number of angina episodes, three heart attacks, two bypass operations, and five angioplasties—the procedure to clear the arteries by reaming them out. He has been on the medical cutting edge for years; he was one of the first heart patients who was treated by laser angioplasty, and some of the drugs he was taking were still in clinical trials. Nonetheless, his condition slowly deteriorated.

During the first few months of 1999, Charles was having 15 to 20 angina attacks each day. He was spending 16 to 17 hours a day in bed. He would get up to bathe and shave in the morning, and by the time he had taken a shower he would need to go and lie down again. He would get back up and shave, and then have to go and rest again. If he walked downstairs, he had chest pain and had to sit down and rest. If he walked upstairs, he had chest pain and shortness of breath and would have to go to bed. As if his exhausted body was not enough to cope with, the medication was making Charles mentally exhausted. He was finding it difficult to concentrate on even the most mundane tasks. He would try to read the newspaper, but sometimes would have to read the same paragraph three or four times before he understood what it said.

The worst part, he says, was that he felt like he had run out of options. After his latest laser angioplasty failed to make a difference, Charles's doctor had a few succinct words of advice: "You've got to sit still until technology catches up with you."

And finally it did. In April 1999, Charles joined another experimental study, a trial of a treatment aimed at growing brand-new arteries to bypass the failed vessels around his heart. The inspiration for this treatment came, not unnaturally, from the notion of the embryo's regenerative powers. Early in an embryo's development, a local hormone called VEGF, or Vascular Endothelial Growth Factor, is switched on to orchestrate the growing of blood vessels; in fact, this factor conducts the development of the entire circulatory system.

Like most growth factors, VEGF is at its most effective in the earliest stages of life. But VEGF is never shut down entirely. When we cut a finger, the growth factor promotes the formation of new blood vessels to allow scar tissue to be built and, later, to keep it alive. Cancer researchers have had their eye on it for at least the last decade. Ironically, they are trying to work out ways to block its action, and thus prevent tumors from establishing a blood supply or creating more vessels to feed the cancer's cells, which allow them to grow and spread. (See Chapter 4.)

It didn't take long for researchers in another area—cardiologists who spend most of their days working with patients desperately in need of new, clean blood vessels—to start pursuing VEGF.

Jeffrey Isner, a cardiologist at St. Elizabeth's Medical Center, Boston, got very excited when he first considered the potential that this substance might have. He knew that if it were possible to use VEGF to make blood vessels grow, then maybe entirely new blood vessels could actually be created in areas of the body that were not getting enough blood flow—without surgery, and without angioplasty.

Although VEGF can easily create an entire circulatory system from scratch in the embryo, it may well struggle to provide a cholesterol-clogged adult with enough clean vessels to keep the heart muscle healthy and pumping. In some patients, the process works perfectly; some people are able to tolerate blockages of blood vessels because they seem to have a great ability to turn on, or "express" VEGF, and form those blood vessels on their own. Dr. Isner came to believe that the patients he is treating are patients who generally do not have that ability. For some reason, these patients could not express VEGF normally.

So Isner decided to do it for them. He decided on a form of gene therapy. By providing these patients with more and more copies of the VEGF gene, he hoped to cajole their vessel-producing systems into action. In order to get the gene into the body, the team had to inject it directly into the heart.

Gene therapy is potentially very dangerous. There is a danger of causing mayhem if the injected DNA disrupts the actions of other essential genes. And even cancer may be started if certain genes are disrupted. Another serious problem is that genes injected in this way usually only work (or "express") for a few days, perhaps two weeks. That, you would think, is certainly not long enough to overcome a lifetime's damage.

The VEGF trials began with desperately ill patients who were facing leg amputations. The blood vessels in their legs were so blocked up with thick yellow plaques of atheroma that their circulation was nearly nonexistent. Originally, the doctors used a catheter to introduce the gene into

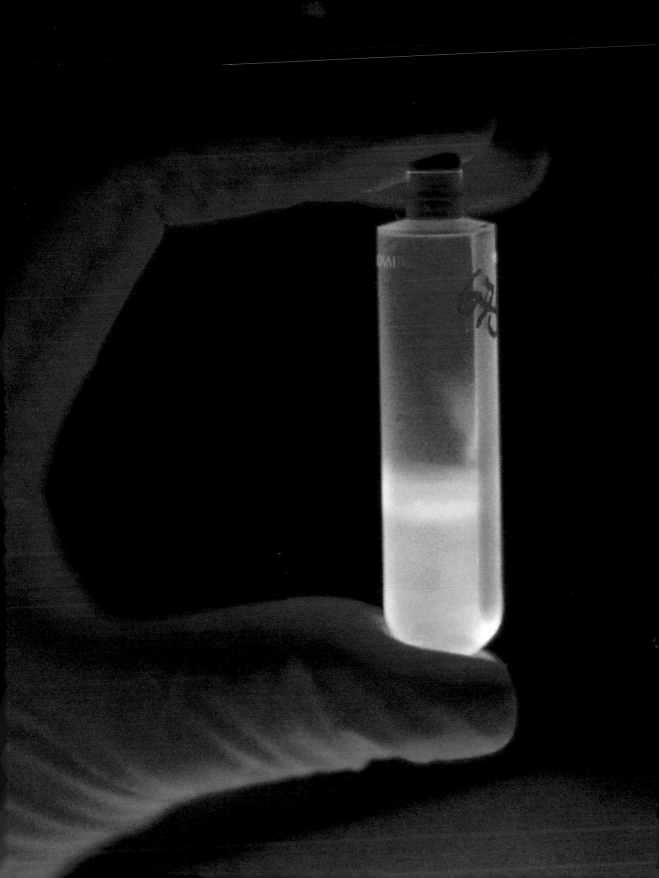

any artery in the leg that could carry them into circulation. But some of the patients' blood vessels were so far gone, says Dr. Isner, that it was impossible to find such an artery. And so they began to simply inject the genes directly into the leg muscles. It was a clumsy approach, but it worked.

It only took one patient for them to realize that they had stumbled upon an extremely effective technique. The genes were being put to use more quickly, pumping out much more VEGF than they would have done before. Isner didn't bother using catheters anymore.

But Reverend Wilson needed more VEGF in his heart, not his legs. However, when Isner and his team was finally ready to move from legs to hearts, they ran into a bit of a problem. They now knew that direct injection was the best approach, but doing injections directly into the heart is difficult and, if you can't see where you are going, potentially dangerous. In the end, they decided to make a very small, so-called keyhole incision in the chest, an incision large enough to allow access to the heart for reasonably precise injection of the genes, but small enough that the patient did not need to go on a heart-lung bypass machine.

On April 22, 1999, technology at last caught up with Reverend Charles Wilson, when Isner and his team injected VEGF-making genes into the muscle of his heart.

By all accounts, Charles's new blood vessels are growing beautifully. Three months after the procedure, despite becoming more and more active, he is not experiencing any angina at all. He can take a shower without feeling chest pain and goes out for a daily walk at his local mall. Four days a week he works out on a stationary exercise bike for 20 minutes. Of course, nobody knows how long the effect will last, but for the moment the treatment has been a stunning success.

To look at Reverend Wilson's body is to read the history of his disease. All you have to do is follow the trails of scars: "Well, this one running down the center of my chest—that nice long one—that's where we've opened my chest twice, just like you open up the hood of a car, so the doctor can get his hands in there and work on the heart," he says, beginning the tour. "I have scars that run from the ankle to the hip bone on each leg, where they have opened up the leg and taken out a blood vessel that would bypass the blockage in the coronary artery. And then the short round one here is where they opened my ribs and injected the DNA into my heart."

Charles remembers the day after his treatment when he realized the angina was beaten.

OPPOSITE *This vial contains the modified AIDS virus, HIV. In this form it can be used as a viral vector to piggyback new genes into cells for gene therapy.*

"One day as I was walking through the mall, I turned to my wife, who always walks with me, and said, 'I don't have any pressure or pain or anything in my chest.' And we had walked about a quarter of a mile. Before the operation I couldn't walk from my house to the car without chest pain. Now, I'm walking one and a quarter miles."

Reverend Wilson was the 31st patient to undergo the VEGF gene therapy. Roger Darke became the 35th, just three days after the 35th anniversary of his marriage to his wife, Susan. He had undergone a quintuple bypass operation in November 1993. The bypass grafted new vessels across his heart in a three-hour operation, but four of the five bypasses had subsequently closed up. Soon the resulting angina was really incapacitating him. He could no longer go bowling or do lengths in his swimming pool; in fact, he couldn't even skim the leaves off its surface without precipitating an attack.

"I have three little granddaughters under two, and I can't carry them around," he said, just a few days before the gene therapy procedure. "I have some little grandsons—five and six—and I can't go out and throw the ball back and forth with them."

Roger Darke could no longer be considered as a candidate for bypass surgery because of his failed grafts. Nor was he really suitable as a heart transplant candidate because his heart muscle was thus far unscarred by his failing circulatory system. It just caused so much pain that his activity was completely limited. So Roger came to Dr. Isner by referral from his own cardiologist. He was, by all accounts, a perfect candidate for the gene therapy trials.

Roger's operation on the morning of May 25, 1999, went smoothly. Isner was happy as he left the operating room: "There weren't as many adhesions between the surface of the heart and the pericardium (the lining) around the heart as there are in most of these patients," Isner reported, "so this one was remarkably simple from that perspective. Everything went very well. We were able to deliver the gene exactly where we want it, so I would say there's a chance of a good outcome."

The anesthesia had really knocked Roger out. He kept complaining that he could not keep his eyes open, kept drifting off to sleep when people came to visit him. That evening, the doctors suggested that Susan go home and get some rest. She was reluctant to leave, but decided to let Roger sleep it off. She left sometime after 8 PM.

Around 6:15 AM, her phone rang. In those early morning hours Roger had complained of some chest pain, then his blood pressure had plummeted. Over the next hour and a half, a team of physicians and nurses tried to raise his blood pressure; each time they succeeded, however, his heartbeat would become unstable. He developed an arrhythmia, a change in the normal rhythm of the heartbeat. Some 90 minutes after complaining about chest pain, Roger Darke was dead. An autopsy revealed nothing remarkable—indeed, it revealed nothing useful at all.

Dr. Isner reported the death to the US Food and Drug Administration (FDA) as he was required to do. The FDA examined both the autopsy report and some of the heart tissue. This was a nervous time; trials can be—and have been—stopped in their tracks if a patient dies for unexplained reasons. But the FDA gave Isner the green light to continue with the trial. After several days, they concluded that this appeared to be a function more of his own disease than of the gene therapy. After all, Roger's death had occurred within hours of the treatment, well before the newly injected gene could have produced any effect.

Susan Darke is left with a long list of unanswered questions. "I guess the biggest question for me has been, would this have happened anyway? Was this waiting to happen? The doctors feel that it was." However, Susan is philosophical about the treatment. She is not left with any bitterness over the trial, or Roger's decision to take part in it. "His living was beginning to stop," she says simply. "You know, for 35 years," she adds with a sad smile, "he let me make every decision but this one. And there have been many times now when I've been relieved that it was his decision. I want to feel good about his choice, because he believed in it."

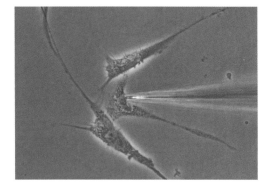

Genes can occasionally be injected directly into individual cells. Here a lymphocyte is being injected using a fine glass needle made in the laboratory. The tip of the needle is completely invisible to the naked eye.

So far, the death of Roger Darke is the only tragedy associated with Isner's otherwise successful trial. But, not long after Roger's death, a patient, Jesse Gelsinger, died in an entirely different gene therapy trial at the University of Pennsylvania. It was thought this patient's death was caused by the adenovirus used to deliver the gene. Within a few days of the viral injection, all the blood in his veins and arteries slowly turned to sludge. The reason why a virus was used is because this is the most efficient way of transporting new DNA into the cell nucleus. After isolating a suitable virus in the laboratoryoratory, part of its own DNA is snipped out. This part is replaced with the gene sequence that the doctors transfer to the patient. Once enough modified virus has been grown in the laboratory, it can be injected into the patient. Hopefully it will then infect the patient's cells. Once the infection has happened the virus should release the desired DNA into the cells, where it will start to work. But viruses are potentially dangerous and much can go wrong. Dr. Isner's trial was closed down in response to Jesse Gelsinger's death. It is, potentially, a huge loss.

Dr. Isner has estimated that, if proven effective and approved, his technique could help up to 250,000 patients a year. But the Pennsylvania death has raised questions over the use of gene therapy, and the delivery system for the genes must be proven safe before trials can continue.

Isner and his colleagues are confident that these hurdles will be overcome. But use of these growth factors and other kinds of gene therapy are limited. Gene therapy may work for growing new blood vessels—essentially all it takes is the manipulation of a single gene—but it is not capable of growing a new arm or a new kidney; these are made by an enormously complex network of genes and growth factors in a process that is barely understood. New blood vessels can be grown in just a couple of weeks. But to grow a new kidney, or even, for that matter, a new heart, we would need to gain control of many, many more genes for a much longer period of time.

> … gene therapists cannot grow you a new kidney … You would need a human tinkerer, someone with the audacity to try to create bits and pieces of human beings in a laboratory dish or incubator.

In other words, gene therapists cannot grow you a new kidney, no matter how good they are at their craft. For that, you would need an engineer. You would need a human tinkerer, someone with the audacity to try to create bits and pieces of human beings in a laboratory dish or an incubator.

The first steps toward that strange scenario have begun, and the first body part to be grown outside the body was skin. Today, bioengineered skin is big business. Live skin grafts, the first fruit of the tissue engineering field to reach the market, are currently supporting a multi-million dollar industry. There are the so-called dermal grafts, made only of the middle layers of skin, and, more recently, the Apligraf, the first substance to have all the components of natural skin, including both the dermis and epidermis.

Apligraf's tiny patches, three inches in diameter, are currently used to treat venous leg ulcers, which are the result of defects in the veins around

the ankle, and diabetic leg ulcers. The patches are derived from, of all things, human foreskins. Why? For one thing, yards upon yards of foreskin are removed from infant boys every year, voluntarily, by routine circumcision. The company needs only to get the mothers' permission to use that which is normally discarded. And, with careful culturing, each of these foreskins can yield as many as 200,000 grafts—enough skin to cover six football fields.

Foreskin cells are ideal. They are young and vigorous, and because they are more or less immunologically naïve, they are unlikely to provoke too much of an immune response. But, just to be sure, the company skims off the immune cells from the rest before beginning the engineering process. The remaining cells are then separated into fibroblasts (the tough, fibrous forerunners of the dermis, the lowest layer of skin) and keratinocytes (the cells that produce the skin's top layer, the epidermis).

The fibroblasts are grown first, along with some added collagen, which is the most prevalent protein found in skin. The fibroblasts arrange the collagen fibers into an ordered matrix upon which they can grow. Then, once the fibroblasts have proliferated and formed a dermal layer, a layer of keratinocytes is placed on top of it. The layers are fed three times a day with a perfect mix of nutrients, carbon dioxide, and oxygen to coax them into creating an epidermis. Once that growth is complete, the cells are exposed to air, prompting them to form the outer, protective layer of skin that is called the stratum corneum.

In less than two weeks, the tissue-engineered skin is ready for grafting.

Making skin, however, is relatively easy. The real challenge, say the tissue engineers, is building a living, working organ. But there are immensely complex problems. It requires an understanding of how the body gets individual cells to cooperate with other cells to form living, growing tissue. It requires a complete comprehension of how the body feeds and maintains those cells; we need a near-perfect reproduction of that internal catering service. It is an attempt to mimic the extraordinary feats performed in the embryo, the cell factory in which all of our limbs, organs, and tissues are manufactured.

Charles Vacanti is Director of the Center for Tissue Engineering at the University of Massachusetts, and, along with his brother Jay, was one of the first of the audacious tinkerers. (There are eight Vacanti children, and, today, four of them are involved in tissue engineering.) They say their work is similar to what physicians have done for hundreds of years—to create the conditions in which the body can heal itself. This is

Superhuman medicine at its most ambitious. They want to push the body's own powers of recovery and regeneration to the extreme, retooling the factory so production can begin again.

The brothers Vacanti may be forever known as the men who grew an ear on the back of a mouse. Most people will recall the *frisson* of horror that the photograph of this mouse caused when it was featured on the front pages of newspapers. Some people believe that this photograph did considerable damage. Certainly, many ordinary people were revolted at a picture of a mouse with this huge ear—like a sail—arising from the middle of its back.

> Making skin … is relatively easy. The real challenge, say the tissue-engineers, is building a living, working organ.

However, this is not something the Vacanti brothers say they are ashamed of. After all, it was a powerful demonstration that several years' work, which involved trying to grow living tissue on an artificial scaffold, was not only fruitful in the laboratoryoratory, but clearly possible in living creatures as well. Some people maintain that the cartilage grown in this way was too flimsy to ever be used to reconstruct a human's ear. Whatever the truth, the story got turned into something much less about science, and more about journalistic outrage.

The late 1980s were frustrating years for Jay Vacanti. He was a reconstructive surgeon and often performed liver transplants on children, but felt stymied by the lack of donor livers availaboratoryle. Jay had seen some of the earliest attempts at creating artificial skin for burn patients and began to wonder whether the same might be done for other organs. It occurred to him that that was an approach one could use for building living tissue. He thought, If we can't find the donor livers, maybe we can build them.

Jay's thinking was not just unorthodox, it was fantastical. Charles says that Jay's ideas were not taken seriously. "I think the fact of the matter was it had never even occurred to most people. So it wasn't that people walked around thinking, 'It's not possible to grow an organ.' I think it's more that people didn't even consider it. Jay was one of the first people who not only envisioned that it was feasible, but also prospectively developed a practical approach that, when you think about it, made a lot of sense and made people stop and think, 'This may indeed be possible.'"

Jay and his colleague, Robert Langer, realized that the best way to grow organs would be, first, to learn how to grow the individual tissues out of which an organ is made. And, in order to do that, they would need to start with the most basic of living units—the single cell. Once again, the answer lay in the very first stages of human life.

In a developing embryo, cells rely on a series of different cues—hormonal, spatial, temporal—to determine what they are, what they should become, where they should go, and what they should do. To imitate this in an adult body, the scientists knew they would have to build some kind of physical structure—a kind of scaffolding—in order to give the cells their structural cues. And it would have to be a biodegradable scaffolding, so that, once the cells had organized themselves properly and were ready to set out on their own, their artificial prop would disappear.

What sort of structure should the artificial edifice have? What sort of structure would allow each and every cell to be fed and oxygenated and given all the hormonal information it would need? Those questions quickly became Jay Vacanti's constant companion; the answer came to him at the beach. "There was seaweed in the water, just waving back and forth. And so it finally caught my eye that that was the geometry that solved

The controversial mouse experiment. Here the loose skin on the back of the mouse acts as a vehicle for the growth of new human cartilage to make a prototype human ear.

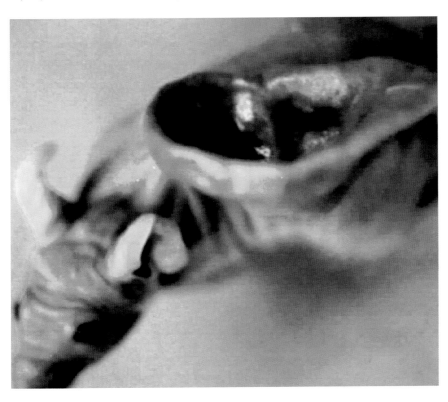

the problem. No matter how large the seaweed was in three dimensions, as it waved through the water the inside was no worse off than the outside because of the large surface area."

Seaweed is in no way unique, structurally—indeed, a number of biological creatures and many of their organs share its elaboratoryorate branching form. The human lung, for instance, with its delicate filigree of ever-smaller air tubes and blood vessels, is a masterpiece of branching. Each cell of the lung is never more than a layer or two away from an air space or a capillary. This kind of structure turned out to be perfect for nurturing delicate cells. But there were still significant problems to overcome. For one thing, the liver cells that Jay was working with were so oxygen-hungry that he had great difficulty keeping them alive and properly "fed."

That was when Charles stepped in and suggested a change of plan. He had been working with cartilage, and knew that cartilage cells do not require much oxygen at all. Together, the brothers conducted an experiment with cartilage cells; they seeded them onto a polymer scaffolding and kept them in an incubator for several days. The cells, it turned out, were very happy in their new home. They stuck to the scaffolding, multiplied and grew—grew into the exact shape of the structure provided.

The Vacantis were thrilled; cartilage does indeed grow on trees. However, when they published their cartilage experiments in a journal, the scientists who read it were underwhelmed. They said to the Vacantis, this is nice, but so what? Can you make specific shapes? And so the idea of the mouse with the ear was born.

The mouse with the ear on its back was a good experiment, both men insist. It did precisely what it was supposed to do: It demonstrated that not only could cartilage be grown on these polymer scaffolds, but it could be grown in a living creature rather than a petri dish. It showed that, even when grown in a living creature, the scaffolding could coax the cells into a very precise shape, a shape more perfect and precise than any plastic surgeon would be able to craft.

If it was such a success, why then do both men squirm when the experiment is mentioned? The problem, according to Charles, is the way in which its results were interpreted—or misinterpreted. "I've read that it was a genetically manipulated mouse," he sighs, "that the ear grew on the mouse and we were going to transplant it into humans."

Not so. The intent was to demonstrate that the technology was possible. In other words, if an ear implant was ever to be used in humans, the plastic structure would first be seeded with the patient's own cells, then implanted into that same patient. Once the plastic inside had broken down, the result would be a perfect ear made of the patient's own cartilage cells and grown into a particular predesigned shape.

The brothers would like to be remembered as the men who learned how to create and shape cartilage and bone. As we shall see later in this chapter, they are now experimenting with regrowing broken nerve connections, and, perhaps, would like to be remembered as the men who discovered how to regrow a spinal cord. That may still come to pass, but, for now, the mouse ear looms large.

Anthony Atala of the Harvard Medical School sees himself as a sort of high-tech baker in the world of self-repair. "Every step counts in terms of making the perfect cake," he says. "And every step counts in terms of making the perfect organ."

Dr. Atala's recipe for creating an organ from scratch sounds simple. Obtain a biopsy from the patient who will eventually receive the new organ. Divide the cells into their individual components. Grow each of these cell types out for about four weeks, during which time you will get enough cells to cover a football field. Create a mold of the three-dimensional structure of the organ in question. Layer the cells over the mold, one layer at a time. After each layer is in place, put the mold in an incubator and bake at 98.6° Fahrenheit for about 24 hours. Repeat for each cell layer, until the organ is finished. And then, says Dr. Atala, implant carefully into the patient.

In reality, of course, the process is almost unbearably complex. It took Dr. Atala nine years to bake a working bladder—the first true organ ever grown outside the body.

The bladder is not a glamorous organ. Unimpressive at first glance, it is nothing more than a flaccid little sac made up of thin layers of smooth muscle cells, lined with a mucous membrane, and held together by an outer coating of tough, fibrous connective tissue. But hook it up to a couple of kidneys and it can perform some remarkable feats. Those layers of muscle crisscross one another in such a way that allows this sac to expand to several times its normal size and to hold up to two pints of urine. We filmed my drinking in a pub, to demonstrate this very point. The bladder can hold that urine in until the the sac is relatively full, and then squeeze every last drop of it out through the urethra; it does this four to six times a day.

In recreating this masterpiece of evolutionary engineering, Dr. Atala came up against a number of stumbling blocks. For one thing, he had to work out exactly what chemicals to use to coax the three different types of cells to divide and grow. And then he had to find a material for the mold that would allow those cells not only to thrive but to organize themselves into a working organ. After all, a bladder that cannot expand or contract properly, simply will not, so to speak, hold water.

Dr. Atala's bladders have been working perfectly in animals, some for close to a year. He has implanted 50 bladders in animals, and he has also been baking some tracheas and kidneys, still in the earlier phases of testing. Soon Dr. Atala may get approval from the US Food and Drug Administration to cook up a bladder for a human being.

A man in the pub holding a bladderlike balloon. The bladder can hold up to two pints of urine at a time.

Atala and the rest of the regeneration *avant garde* are delving further and further back into our developmental history. They are seeking to learn the tricks of human development from the embryo, in the hope that they can recreate them artificially. This quest has taken them to the root of every single organ in the body, and a few select and powerful cells, called stem cells.

The stem cell is the most primitive of cells in any given organ. And yet, they have all the power and all the potential. It is from this type of cell that all other cells in the tissues are made; later in life, it is from this cell that all other cells in the tissue are replaced. It is the progenitor, the parent, of the differentiated and specialized cells that carry out the thousands of different functions in each of our organs.

Stem cells are like the stem of the lineage, the same way we talk about the stem of a flower. They are the most primordial base cells in any organ.

When Evan Snyder, a pediatric neurologist at Boston's Children's Hospital, talks about his work with the brain and its own set of stem cells, he favors botanical metaphors. "Maybe you had a beautiful lawn," he says, "but then the weather has been horrible, and maybe you have kids who were riding their bikes back and forth across it and just trashed one part of the lawn. Or maybe you had bad seeds to begin with. You tried to get your lawn going, but you just had a defective product. What you may want to do is reseed the lawn and start over again."

OPPOSITE These cells are among the most exciting discoveries of the century. They are stem cells and lead to the formation of new blood. Stem cells are the very foundation of life.

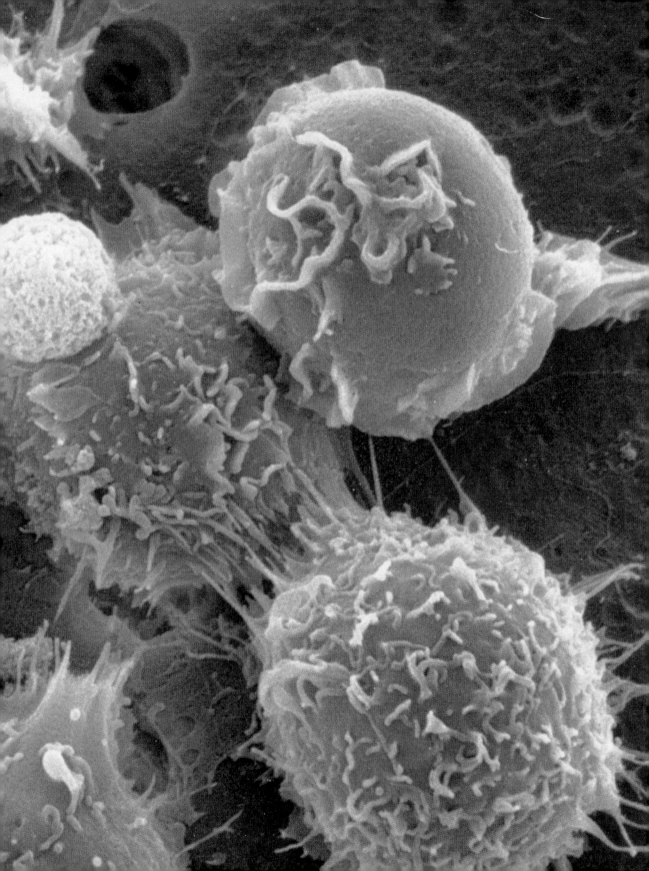

He says the same goes for the human brain. When a seizure, a stroke, or a congenital defect kills off part of the brain, the best solution of all would be to simply start over again, to reseed the brain; thanks to the work of Snyder and a handful of other neuroscientists, that may well be possible in the future.

But stem cells are not just like any old seeds. Snyder says we should imagine seeds that have built-in intelligence. Not only do they give you grass when they hit the lawn, they are so clever that when they fall into the flower bed, they know how to grow tulips. The same seeds can grow grass, flowers, tomatoes, or oak trees, depending on where they land.

Dr. Snyder is convinced that brain stem cells are a kind of "intelligent" superseeds. He thinks they can recognize the part of the nervous system in which they are put, and will begin regenerating tissue accordingly. He wonders if we could reseed the brain or the spinal cord and start these parts growing again with fresh young cells.

The newborn brain is tremendously resilient. Joey is a lively, happy, two-year-old. His mental development appears to be completely normal; he can walk, and he will soon be able to talk. But when he was just two months old, a stroke knocked out a large portion of his brain.

An injury that size in an adult would have put him or her in a wheelchair, and probably taken away the power of speech. Joey's brain, remarkably, has recovered. On the scans, it is possible to see the damaged area but most of the space has been filled in with new tissue. No one realized that the young brain was capable of such accomplished regeneration.

However, can this plasticity be conserved in the adult brain? The conventional thinking was that mature brain cells had traded the ability to bounce back after injury for their remarkable complexity and diversity, and their ability to make hardwired connections. Once a brain cell took on a structure and identity, the thinking went, it never looked back.

It used to be thought that the nervous system was constructed according to what Snyder calls the British plan. "In other words, you are determined by what your ancestors were. You know if you were a lord, you came from lords. If you were a commoner, you came from commoners." In the brain, this means that if you were born of a neuron, you would become a neuron. If your parents were oligodendrocytes, you, too, would be nothing more than an oligodendrocyte.

As a young postdoc, Snyder became intrigued by this elaborate class structure. He began by separating out different young cells from the brain and growing them in culture. His aim was to put them back together and determine how the brain cells organize themselves and communicate with one another.

The problem was that, every time he isolated a cell and grew it in a dish, he would finish up with a mixed population. No matter how often he tried, he could never, ever get a pure population, even when he started with a single cell. He would see all kinds of different cells growing, as if the brain were recreating itself in front of his eyes. The young scientist was intrigued, but the experiment was deemed a failure. Still, even after he was put on an entirely different project, the renegade brain cells kept nagging at him.

The question was, would these cells survive in a living brain? Snyder began to sneak out of the house and into the laboratory at night to continue the experiments. He marked the cells with a blue dye and transplanted them into the developing brains of very young mice. Many months later, it was time to look at the results. It was two in the morning, the laboratory was deserted. Snyder pulled out the frozen sections of mouse brain into which the cells had been implanted and popped them under the microscope. He could not believe his eyes.

The slide filled with blue cells, far more cells than he had implanted into the mouse brain. These cells were perfectly integrated with the rest of the brain. He says his eyes almost popped out of his head. It was one of the most exciting things he had ever experienced in his life.

Overwhelmed by the potential significance of what he was seeing, Snyder took the slide out of the microscope. "I looked at it from a distance, and, sure enough, there was this little blue streak. And, in fact, I actually thought, well, maybe I just by accident marked it with my pen, my blue pen, and this isn't real. So I started looking at other slides, and slide after slide would have this blue streak—this blue streak that you can even see with the naked eye. And then I took many of these and started putting them under the scope, and section after section after section would have these blue cells completely filling up a normal layer of the brain, a region of that had developed when I put the cells in nine months before. And now here they were, completely normal."

"My heart was just racing," he continues. "Because in one fell swoop, in seeing this, I realized what the potential of this phenomenon might be. It meant that we could have a single cell that we could grow and grow and grow in a tissue culture dish, that could be very plastic."

Today, the cells that Snyder found, those little blue cells that curved across the glass slides that night, are known to be neural stem cells. The discovery of the neural stem cells means that the entire scientific community has to throw out their idea of the "British-plan" brain, and embrace an American model.

Dr. Snyder says, patriotically, "The American plan is much more plastic. You can be whatever you want to be. You can go to any part of the country, and if you work hard and study hard enough, you can become President of the United States, or you can become an actor, or an athlete. Now, conversely, the fact that you have incredible athletic ability means that certainly you *can* become a professional athlete, but it does not mean that's what you are *going to* become. The environment will determine whether you really realize your potential."

In other words, most if not all of the 9,000 different nerve cell types estimated to exist in the brain are descendants of the neural stem cells, which migrate to a certain area in the brain and then change into the type of nerve cell that is required. Contrary to expectations, the American dream is alive and well in the supposedly class-ridden brain. The brain, to everyone's surprise, has *plasticity*.

Playing with this plasticity—finding out just how far a stem cell can stretch—has filled most of Snyder's days since his initial discovery. He began by watching the cells in a culture dish, seeing how they divided and differentiated. In fact, he says, the cells not only did an accomplished job of dividing and differentiating, they even interacted with one another. He would see development taking place in the dish right in front of his eyes. But he knew that a tissue culture in vitro does not reflect what goes on in the real brain; all kinds of unexpected things can happen. The logical next experiment was to take these cells and put them back into a developing brain.

The neural stem cells felt completely at home. Implanted into a living mouse, they are remarkably adaptable and seem to know exactly what their new brain needs. Put them in a developing brain, and they start producing neurons. Implant them in a normal adult brain, and they become support cells. But their full potential is realized when they are put into a damaged adult brain.

Neural stem cells implanted in a damaged adult mouse brain will produce new neurons to replace those that are missing or missing in action. They will also produce any and all of the support cells needed to allow those neurons to work properly. These same stem cells will even do a little traveling, if necessary. Stem cells placed in an undamaged portion of the brain will seek out the damaged part, bed down, and begin their repair work—even if they have to make their way all the way to the other hemisphere.

This is the beauty of the nervous system and the beauty of these stem cells. In many cases, particularly at an early age, the system works perfectly without our intervention. There seems to be a very beautiful, intricate system of communication between the cells and the brain as a whole; in fact, the brain sends out signals that order the stem cells to repair it.

But at times it may need some extra instruction. Although a damaged brain will almost always rouse its own stem cells in an attempt to repair itself, in an adult brain that level of activity is normally very, very low, well below what would be needed to stitch up the damaged area properly and restore its function.

But Dr. Snyder says that one strategy to repair brain damage may be to augment this self-repair impulse by implanting even more stem cells. The stem cells could then, potentially, do more than repopulate a damaged area of the brain. They could eventually be used to regrow the nerve cells in the spinal cord, perhaps even reversing paralysis. They could be turned into delivery vehicles—ferrying growth factors to boost the brain's own production of cells and enzymes, bringing in key neurotransmitters or other chemicals that might be depleted from the brain, even transporting genes across the normally impassable blood-brain barrier. Their peripatetic nature could be exploited even further: They could be used to chase and destroy equally mobile cancer cells.

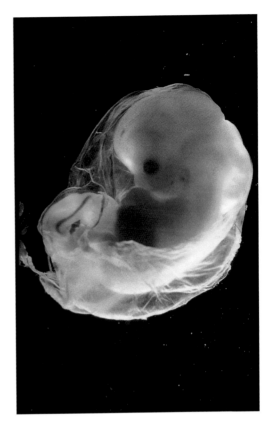

A potential but controversial source of stem cells for treating brain injury and Parkinson's disease victims. So many early fetuses are lost during development that many people believe that using cells taken from them may be justified ethically.

Dr. Snyder is bullish about the possibilities. Alzheimer's disease, Parkinson's disease, brain tumors—all are terrible scourges. He says there is hardly a disease of the brain that could not be addressed by simply sending in a stem cell. But that way of thinking, says Snyder, is narrow minded. The real beauty of stem cells, he insists, is that we may no longer need to try and come up with a different approach for each individual disease and medical problem—perhaps a brain that has suffered a stroke, been hit by a hammer, ripped apart by a bullet, or infected by a virus. "Wouldn't you like to undo all that damage and go back to the way things were before all that happened?" asks Snyder. "Here's a way to do it. Here's a cell that'll allow you to do development all over again."

Whereas many other researchers in the field would be skeptical, Snyder is entirely confident. But at least one experiment has demonstrated the concrete possibilities of brain regeneration – the case of the shivering mice.

Shiverer mice are what Evan Snyder calls "experiments of nature." They are mice born with a natural, congenital defect similar in its effect to multiple sclerosis in humans. My good friend, Dr. Carol Readhead, with whom I did the work on sperm described in the chapter on fertility, was the scientist who explored this gene defect by artificially creating a mouse that carried the gene. She was then able to repair the defect, on one of the first gene therapy models and a very important experiment.

Shiverer mice have a genetic mutation in the cells that produce myelin, the insulation around the axons, the long conducting fibers that protrude from the bodies of nerve cells and carry the signals. Without this insulation, the fibers experience shorts, and the electrical impulse that normally should flow smoothly down the axon moves in fits and starts. The net result is that, instead of calmly walking, sniffing, and eating like most other mice, an affected mouse shivers constantly and uncontrollably.

Enter the neural stem cell. Snyder implanted stem cells in the brains of shiverer mice. The stem cells multiplied, distributed themselves throughout the brain, and became myelin-producing cells.

The shiverer mouse photographed by Carol Redhead. It was one of the first, and very important, models for gene therapy.

These cells produced myelin until the axons were properly sheathed. The shiverer mice stopped shivering. In some cases, when the cells were put into young enough mice, the shivering never even started. Now, the mice are wandering happily around their metal and lucite cages, eating, grooming, even doing a few physical therapy laps in a tiny pool.

The stem cells were giving rise to cells that not only looked normal, not only seemed to be making the right insulation, but were doing it in a functionally relevant manner. In other words, says Dr. Snyder, this experiment showed that neural stem cells can, indeed, regenerate the brain.

The Vacanti brothers' white rats didn't shiver, but they also didn't walk. Their spinal cords had been severed.

The spinal cord, tucked away inside the bony vertebrae of the spine, is a bundle of millions of twisted fibers. Many of them are single nerve cells a foot and a half long, stretched between the base of the brain and the rest of the body, which carry electrical messages at the speed of about 300 miles per hour. Every sensation we have and every movement we make is channeled through the spinal cord. With a severed spinal cord, the rats' brains had no communication at all with the rest of the body.

How can we fix this complex and delicate system? Since the spinal cord is effectively an extension of the brain, could neural stem cells be the answer? Would they have the built-in "intelligence" to bridge a gap in a spinal cord?

Charles Vacanti and another brother, Martin, used their experience in building cartilage structures. They seeded tiny plastic scaffolds with neural stem cells, and implanted them in the five-millimeter gap in the rats' spinal cords. The plastic was biodegradable so, eventually, after the scaffold had done its work, it would simply dissolve.

The experiment was something of a shot in the dark. The Vacantis could not attempt to reconnect the individual neurons. They simply seeded the broken cord with neural stem cells, and waited to see if anything would happen.

Incredibly, within weeks of their operations, these previously immobile rats were moving their legs. And, within just a few months, some of them were walking, albeit with a little less coordination than a normal rat. When the Vacantis started the project, they would have been delighted to see any sign of neurological recovery—the movement of a toe, or the twitch of a leg muscle. The recovery of the paralyzed rats was completely unexpected.

Even more extraordinary was how the rats recovered. The Vacantis looked at the "healed" spinal cords, and found that the implanted cells were bridging the gap, but the section of spinal cord they were recreating looked nothing like the original. Some of the nerves were growing up; some down; some sideways. They were growing in random directions and would hook up with anything; some would hook up with each other; some with the nerves above the cut; others would hook up with nerves below the cut. It was messy, but it appeared to have worked.

The Vacantis learned that it does not matter which nerves made connections with which nerves. They say the process is like cutting a multistrand telephone cable. If you cut all the thousands of wires in the cable, you can try to painstakingly match the red wire with the red wire, the blue with the blue, and remake all the connections as they were before. Or you could randomly make any connection and then let the central computer figure out what the connection was by calling each number to see who answers.

It appears that this is exactly what the rats' brains may have done. The spine is so complex and there are so many strands that it would be impossible to re-form the original connections. So the rats' spinal cords made random connections, the nerves grew together in seemingly disordered fashion, but their brain worked out what the connections were. The rats learned how to walk; they had created order out of chaos.

In spite of the enormous complexity of the nervous system, again there seemed to be built-in intelligence about the healing process. It seems that the Vacantis may have provided the infrastructure and seeded the growth, but the nervous system itself appears to have designed the new system and made it functional.

The Vacanti experiments are still very controversial. Many very significant scientists believe that their rats do not prove the value of stem cell injections. They say that rats are not a very good model, because they are capable of remarkable neurological repair and are extremely tough creatures. So the jury is still out over the spinally injured rats.

The Vacantis say that they are not daunted, that their experiments are still at an early stage. After the rats, they need to progress to larger mammals, such as dogs or monkeys. Martin Vacanti has plans to use the technique for dogs whose spinal cords have been seriously damaged in accidents. This is a closer approximation to the therapy's eventual use, in that if it works, it could be used for people who had similar and unpredictable spinal injuries, rather than a clean break.

Martin accepts that human trials are at least five years away. At the moment there are far too many unknowns for human trials to begin. We cannot predict what the dangers of randomly wiring up a spinal cord would be. What would happen if the nerves, which carry pain signals, are mixed up with those that allow movement? Could movement become uncontrolled, or the patient end up in constant pain?

In human trials, each injury will be different. The effect of the stem cell therapy will depend on the scar tissue surrounding the spinal cord and the damage caused by bone fragments. In some cases, it may be better to surgically cut out some of the damaged cord. But the Vacantis are very aware that they must not cause more damage than was there to begin with; this is a cardinal rule of medicine. Despite these hurdles, Martin believes that if the trials are successful, the therapy could become approved for general use in ten to fifteen years' time.

We should not underestimate our natural powers of regeneration. Without any help from neural stem cells or plastic scaffolds Penny Roberts, our parachuting nurse, has started to get some feeling back in her legs; she can feel extremes of temperatures in her feet and she knows if her leg is twisted or sitting on a crease. This is remarkable given the severity of her spinal injury.

There are also other cases, where limited feeling or movement has returned after a spinal cord has been crushed or even completely severed. Martin Vacanti believes that our spinal cord actually has a latent supply of stem cells that, after injury, can make limited connections automatically. This may be why Penny has regained some feeling in her legs. However, relying on her own physiological abilities has it limits.

Penny is hopeful, but she's realistic about the chances of mending her crushed spinal cord. She would love to be like one of the Vacanti rats, taking small, shaky steps, but she realizes that the chances of her leaving of her wheelchair and walking are slim. Nonetheless, she is aware that, although she herself may miss out, the solution will not be too far away. "If people really put a lot of energy into research, and if we really crack this now, we can end it with my generation. And people of my son's generation won't have to go through everything I've been through."

Miraculous as it might be, the neural stem cell is only the beginning. Indeed, says Evan Snyder, perhaps the most exciting thing about the neural stem cell is what it promises for the rest of the body.

"We who look at the nervous system and begin to understand the beauty and power of this particular cell wonder whether those who work in other organs could look at our cell and say to themselves, I wonder whether a stem cell like that exists in my organ—in the liver, in the muscle, in the gut, in the skin. If so, this whole concept of the beauty of stem cell biology, its plasticity and its ability to accommodate to environments, could represent a whole paradigm shift as to how we envision being able to repair the damaged organ and the damaged organism."

Individual stem cells from individual organs do, indeed, have the potential to revolutionize medicine. But the definitive stem cell does not belong to any one organ or system. The most primordial of the primordial cells, the stem of all stem cells, is the *embryonic* stem cell. This is the cell that has the potential to create a human body.

The day after an egg is fertilized by a sperm it begins to divide. Those early cells, the first cells of life, can turn their hand to making almost anything the body needs—just about any kind of cell, tissue, or organ. This cell, which has now been isolated in the laboratory, has sparked off a vociferous series of scientific and ethical debates. The problem is not so much what it can do, as where it comes from. Embryonic stem cells, after all, come from embryos. Human embryonic stem cells would mostly come from human embryos that have not been returned to the uterus after in vitro fertilization (IVF). It is obvious why passions have been inflamed.

There is, potentially, a particularly controversial way of manipulating human embryos, and this involves the technology that is employed in cloning. In spite of what the critics say, it does not involve cloning embryos. This might avoid the ethical difficulties in using IVF embryos, but it raises another set of moral complications because it might need human eggs.

The technology that scientists have in mind is called therapeutic cloning. The term "therapeutic cloning" is genuinely a misnomer; it implies making clones of animal or human embryos, which is not what is done, nor is cloning the intention of the scientists. The technique involves growing only identical cells. Identical cell cultures have been grown in laboratories for several decades—the difference is that these identical cells would be produced from embryonic cells, possibly after transplantation of the cell nucleus. The object is, of course, lifesaving—to produce new tissues and organs. Cells grown in this way might eventually solve the human tissue shortage problem, since we might no longer need to rely on well-timed fatalities or altruistic relatives. If we could eventually grow whole organs and not just tissues, these techniques could solve the transplant shortage problem.

One way to accomplish this feat would be to manipulate an egg or an embryonic cell. The nucleus would first be plucked from a human egg cell. It would then be replaced with the nucleus of a cell from the person who needs the transplant. Thus, virtually all the DNA would come from this person's transplanted nucleus, possibly from a cell from the heart, liver, skin, or indeed, from any part of the body. The tissue produced would be similar to the tissue donating the nucleus. If skin was required, then a skin cell nucleus would be donor; if a liver was needed, then a liver cell nucleus.

What is so fascinating is that these tissues would be totally compatible with the recipient's immune system *because the recipient donated the key genes first from a nucleus arising from one of his own cells.* Indeed, these "cloned" tissues would be the patient's own tissues. Physiologically speaking, the patient would be donating an organ to himself or herself, with the help of someone else's empty egg.

We will probably never match the regenerative abilities of the salamander or earthworm. The human body is, after all, an evolutionary compromise—we can't have it all.

"If I had to choose between being a salamander that could regenerate its tail or being a human who couldn't regenerate its arm, I would choose to be a human," says Jay Vacanti. "They certainly have some advantages, but in the end it only makes them a better salamander."

However, the difference is we do not have to settle for our evolutionary quota. The Vacantis and hundreds of other scientists working in the field are intent on making us into a better human—a Superhuman.

> The most primordial of the primordial cells, the stem of all stem cells, is the *embryonic* stem cell. This is the cell that has the potential to create a human body.

4 Cancer

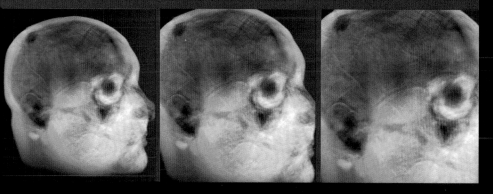

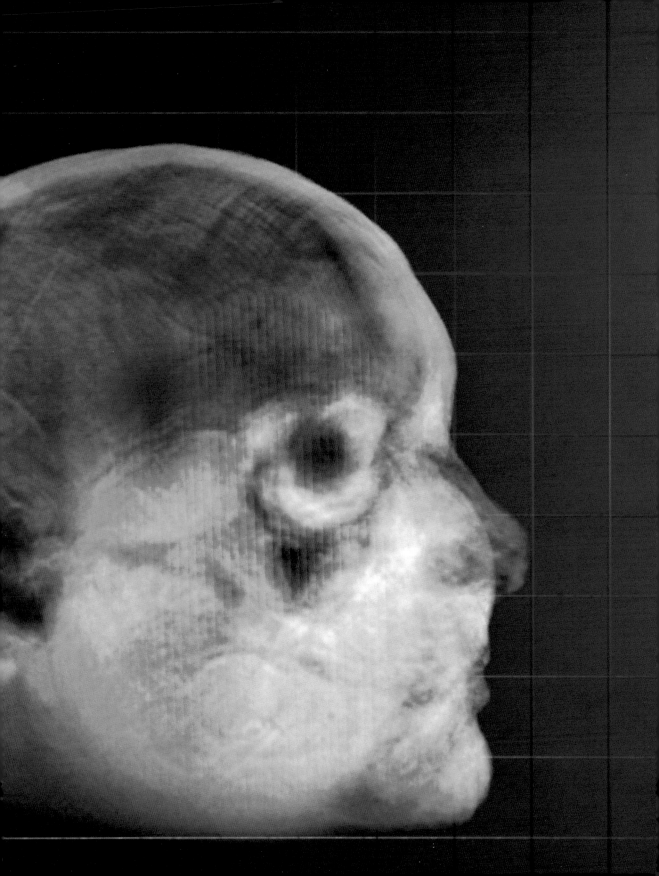

In 1971 Richard Nixon declared war on cancer. He pledged 100 million dollars to fight the disease, then a huge sum. Now, surely, the scourge of cancer would be wiped out. Gleaming new laboratories were constructed and the most brilliant scientific minds were attracted to America's most desirable campuses. A "Cure for Cancer" was duly promised, just like Kennedy had promised to put a man on the moon by the end of the 1960s. If we can put a man on the moon, Nixon said, if we can split the atom, we can crack cancer. This would be the moon shot of medicine, a triumph of man over nature and proof, if proof were needed, that America can accomplish anything it puts its mind to.

Nixon's war on cancer turned out to be a war of attrition. There would be no moment of triumph, no planting of the flag, and no surrender. Across the world that 100 million dollars has turned into billions. The effort has been immense, but there is still no effective, final, absolute cure for most kinds of cancer, a disease that afflicts one in three of the human population.

> Cancer employs guerilla tactics. It can strike quickly … when under attack it can lie low, hide out, and then reemerge stronger and quicker than before …

The United States was not prepared for the Vietcong, and similarly not really prepared for the war against cancer. Cancer employs guerilla tactics. It can strike quickly and often silently; when under attack it can lie low, hide out, and then reemerge stronger and quicker than before; most dangerously, it can change and mutate. Cancers can actually evolve in the body, and like some strains of bacterial infection, become resistant to antibiotics; cancers can become resistant to the drugs that we use to try and kill them.

We are still in the dark when it comes to a great many of the mechanisms and tricks that cancerous cells have up their sleeves. We now recognize that cancer is not one disease. It cannot have one single magical cure; it is around 200 different diseases, and each has its own strengths and idiosyncrasies. The persuasive idea that there will be one cure for the one disease called "Cancer" has finally been debunked.

However, in truth, the situation is not quite as black as I have painted it. It is true that veteran researchers, who have doggedly worked for the cause for years, freely admit that the war on cancer has been superlatively expensive. They fully recognize that it was never likely to live up to Nixon's gung-ho expectations. Nevertheless, even the most cynical critic accepts that there have been very substantial achievements. The focus on cancer research has been of immense medical importance.

Not only have we started to understand the enemy, we have laid the foundations for its defeat.

In the process, we have gained an important understanding of many aspects of cell biology, and cancer research has given us spin-off information about a whole range of diseases of different organs of the body – about immunology, genetics, and even development and reproduction. We also have a finer and more precise model of how cancers produce their effects. We have a firmer grasp of the basic building blocks of cancers: how they begin, develop, spread, and kill. These discoveries, which depend on understanding mechanisms at the molecular level, have laid the groundwork for future therapies and new methods of prevention.

Our military intelligence, as we shall see, is finally bearing fruit.

So why, when we have this understanding, is cancer such a formidable enemy?

Cancer is mostly about losing control. Usually the body is able to regulate and replace its cells according to a reasonably strict timetable. Each healthy cell in the body has its own natural biological rhythms, the cell cycle, a routine and orderly march through a series of stages; there are periods of cell growth, then comes DNA replication, and then cell division, when the cell splits into two. Most cells in humans go through this process about 50 times. In general, the cells of short-lived animals divide fewer times. Mice, who normally live two or three years, have cells that are programmed to divide around 15 times; the cells of the Galapagos tortoise, which lives for about 170 years, may divide around 110 times. After all cells divide for the final time, chemical signals send out a message for the cells to shut down permanently.

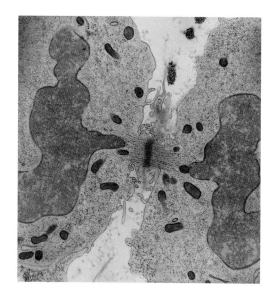

A remarkable photograph of a cell dividing. The two new nuclei are colored green. The spindle, connecting the chromosomes, is in the middle. Cancer cells divide uncontrollably in this way.

The vast majority of cells have a finite life span, whether it is a few days or a few months, but when their time is up they peacefully shuffle off and make way for the new generation. It is literally true that you are not the same person you were even 18 months ago; nearly all the cells in your body have been replaced. This process, whereby cells die naturally, is called "programmed cell death," or apoptosis. "Apoptosis" comes from the Greek and refers to "leaves falling off a tree."

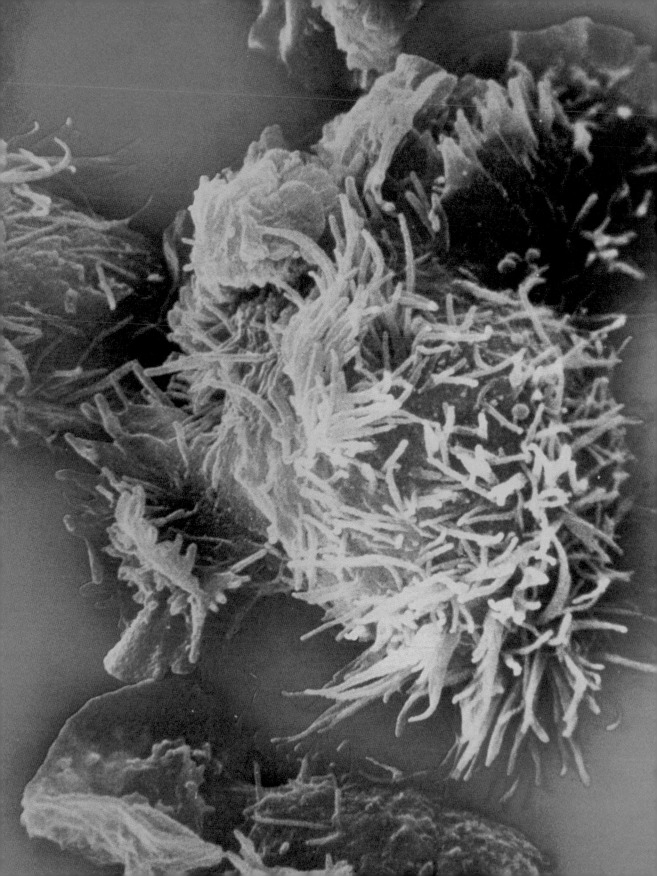

Cancer cells are cells derived from normal tissue, such as skin, muscle, bowel, or bone but a critical part of their DNA has changed. Their genes no longer produce the chemical messages that tell the cell to die, to undergo apoptosis. The cells grow, then they replicate their DNA, and divide. They go through this process again and again and again. They self-replicate indefinitely; they are immortal, and go on producing thousands upon thousands upon millions of identical copies—clones, mutant clones—of themselves. These stubbornly refuse to lie down and die. In most cancers all the millions of cells are descendants—clones—of basically very few cells, the founding fathers of the cancer colony.

Cancer cells do not function normally; they do not act like the tissue from which they are derived; they are not useful. They crowd out, infiltrate, and kill off adjacent healthy tissue. Frequently, they gain access to nearby blood vessels or the lymph system, the network of vessels and glands that runs throughout the body and helps drain and reprocess the fluid surrounding our tissues. Once they gain such access, they can be rapidly circulated and malignant cells can lodge a great distance from the original cancer. Often these distant growths are much more dangerous; they are frequently the growths that result in the death of their victim.

Somehow, these wildly proliferating renegade cells manage not only to evade all cellular control, but to stay out of sight of the immune system. When they break through tissue boundaries, they can sometimes shed their "identity" and this helps in their spread throughout the body. Their initial territorial expansion and movement through the body suggests a stealthy crablike progression, hence the name *cancer*, Latin for crab. Although cancers do not function normally, sometimes they can stimulate the creation of their own blood supply in order to stay well fed, well oxygenated, and alive.

The fundamental irony of cancer is that when these abnormal cells fail to die—when they achieve individual immortality—the body as a whole is put in mortal danger.

Many different factors can alter DNA and predispose cells to grow cancers. There are, for example, hundreds of chemicals that can alter cells in this way. One of the first to be recognized was soot. In 1775, Sir Percival Pott described cancer of the skin of the scrotum in men and in boys after they reached puberty. What they had in common was their employment – chimney sweeping. Their lack of regular washing meant that these carbon deposits had lodged in the skin folds, where they eventually gave rise to cancers.

OPPOSITE A cancer cell typical of those seen in lung cancer. These cells have the property of being mobile and can invade surrounding tissues or migrate to distant sites.

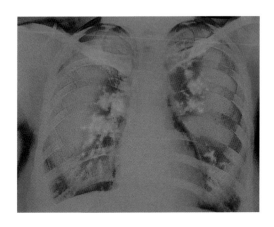

An advanced lung cancer (red) spreading inside the chest. This shocking Xray shows a fatal condition that is largely preventable. Cigarette smoking is a highly dangerous habit.

Remarkably, it wasn't until 1915 that confirmation of this effect was produced in animal experiments. Now, of course, we know that a wide range of hydrocarbons and tars causes cancer, including the tars from cigarette smoking.

Until better health precautions were established, cancers were much more common in workers in the coal and coke industry, in the production of petroleum and oil, and in men who smelted various metals in industry. Other substances that may predispose to cancer are some insecticides—particularly DDT—some compounds previously used as food additives, particular dyes, and mustard gas. Some of the soldiers exposed to poison gas on the Western Front during the First World War subsequently died of cancers.

There are also a number of naturally occurring biological compounds that carry this risk. The most notorious of these is aflatoxin, which grows from the mold *Aspergillus* and which has contaminated foodstuffs in Africa, particularly nuts and cereals. It is thought that the high level of liver cancer in parts of Africa is due to this mold growing in food that is not properly stored.

Chemicals are not the only way DNA can be altered. Another is radiation. Around the turn of the 20th century, it was noted that doctors pioneering Xrays were more likely to develop skin cancer. A significant number of the first people working in radiology—using X rays to diagnose diseases—died of a range of cancers, including leukemia. The inventor of luminous paint, which contained radioactive mesothorium and radium used on watch dials, died of a form of blood cancer in 1928. By the 1930s, it was recognized that women employed in watch factories painting dials were at risk of cancer of the jaw because they licked their paintbrushes between each job to give the brush as fine a point as possible for this delicate work. Since then there has been recognition that cancer might be triggered by exposure to ionizing radiation from a variety of sources, the most worrisome, of course, being nuclear explosions and radioactive waste.

OPPOSITE A contaminant in the food. The mold, Aspergillus, produces the toxin which has caused so much liver cancer in Africa.

Another way the DNA in cells can be changed is by infection with viruses. Numerous viruses are known to cause cancers in some animals, including fish. The evidence in humans is not as good.

However, one of the viruses that causes a herpeslike disease also causes lymphatic cancer, particularly in children—so-called Burkitt's lymphoma. This was first described in eastern and central Africa. Another related virus was described from southern China—it causes a rare cancer of the pharynx, behind the nose passages. Some forms of leukemia seem to be caused by viruses, and it is possible that viruses may be implicated in many cases of cancer of the cervix, one of the more common cancers affecting women.

But to understand cancer, we have to go back to the earliest days of evolution. When we jump back in time a few billion years we come to see that cancer is an inevitable part of our bargain with life on this planet. Two and a half billion years ago the world was very different. Its inhabitants were prokaryotes, single-celled bacteria that lived in an atmosphere composed mainly of carbon dioxide. One theory is that these simple but prolific organisms thrived on this gas. They swam around in nutrient-rich mud and sand, rock pools and puddles, converting water into hydrogen and spare electrons, and releasing carbon from the carbon dioxide in the air to feed their metabolic systems. And this is what they did for millions of years. There was no great hurry for these life-forms to evolve—they were "happy" enough.

But soon they were forced to evolve. The chemical process that uses carbon dioxide as one of its raw ingredients had a by-product—another, far more reactive gas, which was gradually increasing in concentration in the atmosphere. This highly reactive gas was poison for these tiny bugs. It reacted with their proteins and interfered with their entire metabolic process.

The gas was oxygen. The theory goes that as soon as oxygen reached a certain concentration in the atmosphere, perhaps around 20 percent, it wiped out many of these organisms in a bacterial apocalypse. They could not stand the toxicity any longer and their metabolic systems failed. But a few hardy ones survived, probably the few organisms that were living deep under the mud or able to burrow there, and thence escape this poisonous gas. Eventually, this brutal evolutionary pressure gave rise to another, free-living microbe that could actually use oxygen as a fuel.

One current theory, suggested by the American biologist Lynn Margulis, proposes that all complex life on earth, from the amoeba upward, is the result of a happy and long-lasting marriage between these two kinds of primeval organisms—those that hate oxygen, and those that love it.

To understand all this we need to look at the structure of the cell. The cells that make up the bodies of all animals, whether they are pigs, fish, spiders, or humans, are composed of an outer cell membrane that is like a bag inside of which is a semifluid substance called the cytoplasm. A prominent structure in the cytoplasm is the nucleus, which is the command and control center of the cell. It contains the DNA that makes up nearly all the genes that make us what and who we are.

Indeed, until recently, the nucleus was thought to contain *all* the genes. But spaced out in the cytoplasm are many other very tiny bodies. Among the most common of these are the mitochondria, which can be seen under a conventional microscope and whose presence has been known about for a long time. Their structure is complex, and when I was a medical student it wasn't fully understood what they did or what they contained. It turns out that the mitochondria are the "power plants" of the cell, using oxygen and sugar and fat products in making the body's sources of energy.

The DNA in the mitochondria has now been sequenced. Unlike the genes in the nucleus of human cells, the entire mitochondrial DNA is known. This is partly because there isn't very much of it. While the DNA in the nucleus is made up from billions of letters or base-pairs, human mitochondrial DNA contains merely some 16,000. The configuration of this DNA bears a very striking resemblance to that seen in all bacteria.

One evolutionary theory that holds strong sway is that mitochondria are actually descendants of some of the original oxygen-hating microbes which were coming to terms with the new oxygenated world. These microbes burrowed into cells and found a comfortable new home. These tiny critters did not have their own nucleus and cell membrane, so they were safer and better fed; they lived in harmony, in symbiosis, within every one of these cells, metabolizing and producing energy. The combination of these two organisms, the thinking goes, produced a highly successful single-celled organism that was the basis of all single-celled and multicellular animal life on earth.

With the introduction of these microbes, animals developed their own metabolic processes based on those of the bacteria—the reaction that begins with oxygen and food as its raw materials and ends up with energy and carbon dioxide. All animal life in the past couple of billion years depends on this chemical equation and it is this that we human beings have inherited.

These primordial oxygen-hating bugs made their bed inside us, and now we have to lie with them. Some billions of years later, we are feeling the drawbacks of this toxic companionship. During the process of energy production, the mitochondria produce free radicals, highly reactive molecules that are created as a by-product of our metabolic activity.

These molecules are exceedingly unstable; once released they zip away into the body and rattle around, causing a huge amount of damage. For example, it is thought that they can attack the arteries, causing the initial damage that allows fatty deposits called atheroma to stick to the vessel walls and start furring up the arteries. But, worse than that, free radicals damage the DNA inside the nucleus. They are hyperactive molecules that can wreak havoc by causing chunks of the DNA code to be corrupted. They can be considered to be somewhat similar to a computer virus that corrupts the data on a hard disk.

In most cases, the damaged part of the DNA in the nucleus of the cell is unimportant and no harm ensues. Long stretches of base-pairs lie apparently unused because they are not parts of genes. Perhaps they are spare capacity for future evolution, or just redundant code for which there is no use. As it happens, only about nine percent of our total DNA seems to do anything, to be part of the genetic code that gives the formula for the body to make proteins. The rest has been referred to as "junk." Damage to the junk code does not matter one way or the other; it seems to have no discernible effect on cell function or replication.

> ... free radicals damage the DNA inside the nucleus ... they can be considered to be somewhat similar to a computer virus that corrupts the data on a hard disk.

Even if the damage is sustained to a functional part of the DNA code, most of the time the DNA is capable of fixing itself. There are specific genes devoted to repair of the DNA. These send out chemical fast-response teams to zip up and down the double helix to patch up the damage. But, sometimes, it is not completely mended. Sometimes, as we shall discover, there is a mutation in the very gene that is meant to fix the DNA, so that the repair process itself is sabotaged.

Occasionally, then, an important part of the code is damaged and not repaired. This is when the cell can become a mutant. It is changed and potentially cancerous. The DNA in it may no longer "listen" to the signals telling the cell to die; it might go on replicating indefinitely, and that is where trouble starts. The more breaths we take, the more oxygen we process, the more likely we may be to find ourselves under attack from these fierce, immortal, and mutant cells.

This may be why some people have come to believe that a semistarvation diet may allow us to live longer. Certainly, reducing our food intake brings down the body's metabolic rate. Slowing everything down has the effect of reducing our need for, and therefore our intake of, oxygen. In these circumstances all our cells together need less oxygen than normal to function. If our oxygen intake is reduced, then the production of free radicals is reduced and correspondingly this reduces the damage caused by these vicious molecules.

So, the theory goes, eating less not only reduces DNA damage, which reduces the risk of cells mutating into cancer cells, but it also slows down the entire aging process by protecting our cells and tissues from the constant, day-to-day damage inflicted by free radicals.

Certainly the theory seems to have been borne out in mice, which have been placed on diets consisting of two-thirds of their normal food intake (although they are given supplements to make up for the vitamin and mineral deficit). Such mice can live, on average, 30 percent longer than normal. Similar studies on monkeys are being conducted at the time of writing.

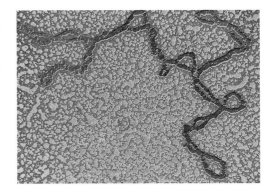

The loop of DNA from a single mitochondrion. Evolution suggests that our mitochondria are derived from bacteria that infected animal cells billions of years ago.

However, starvation has other important side effects; a key one is a reduction in the body's defenses against attack by foreign proteins. It is well known that starvation reduces our resistance to infection, which is, of course, part of the reason why famine is so serious in many of the poorer parts of the world. Starved populations don't only die from lack of food, but are far more likely to suffer from tuberculosis and the ravages that result from HIV. And it is also likely that starvation may reduce human resistance to cancer cells.

However, humans tend to eat more than they probably need and there seems to be an intriguing reason for this. Clearly there could be a better balance between eating too much and starving ourselves into sickness and ill health. You might think that during the course of evolution we would have learned to reduce our food intake and therefore our metabolic rate, because this genetic strategy would on balance prolong our life. But a low-calorie diet has another, unwelcome side effect—it lowers the sex drive. It seems that when humans eat only just enough food, they become more interested in eating than having sex. Who has not put off going to bed to linger over a fabulous chocolate dessert? In the context of evolution that is not a desirable state of affairs. Perhaps the popular image of the sex-hungry early caveman is not so realistic after all.

As it happens, in evolutionary terms, cancer was never really much of an issue. Cancer is a disease that becomes far more prevalent as we get older—the product of a number of factors, some controllable, but others random. The random damage caused to our DNA is like the result of a constant game of Russian roulette played with a gun in which several million barrels are empty and just one is loaded. Free radicals corrupt our DNA on a daily basis, but nearly all the time the damage is either benign, or repaired by the DNA itself.

On the relatively rare occasions a cancerous cell develops, it does not survive and undergo division. But over time, we are under constant bombardment. The more years we spend being exposed to free radicals, the more likely it is that the Law of Averages will make its ugly presence felt and the chamber of the gun will be holding a live cartridge instead of a blank. It is a matter of statistics and this is why cancers are more prevalent in old people than in the young. Stone Age humans almost certainly did not live long enough for cancer to become much of a problem. By the time the free radicals had done their worst, these early humans would have procreated and raised their children, avoiding any evolutionary disadvantage even if they were genetically disposed to cancer.

Stone Age humans had another significant advantage. They were forced to be healthy eaters—meat was almost certainly a relative rarity. In former times, we humans used to get a majority of our food supply from cereals, or if we were lucky, from vegetables and fruit. All these dietary sources have concentrations of chemicals called antioxidants.

These are concentrated particularly in green leaves, which collect carbon dioxide from the atmosphere, react, and neutralize the free radicals. The leaves have to be able to deal independently with oxygen, which to plants is also a kind of poison. Those antioxidants, produced protectively by plants, can also protect man.

There is another mechanism that is fundamentally important to the understanding of cancer. DNA copying is not always perfect. Whenever our cells divide, the entire complement of chromosomes, containing some 100,000 genes, is duplicated. It is a good thing that this process is slightly imperfect, for without this intrinsic mutability there would be no evolution. We would all look the same. Indeed, without it, we would still scratch out an existence as single-celled creatures burrowing in the primeval mud and avoiding oxygen poisoning. However, the process of cell division has the downside of producing potentially malignant mutations, and, as already pointed out, our fragile DNA is also under constant attack from many environmental toxins—naturally occurring geological radioactivity, ultraviolet light from the sun, and, of course, the modern poisons of cigarette smoke, pesticides, and other carcinogenic chemicals.

So cancer, although exacerbated by modern toxins and certainly not helped by our fat-laden diet, is not a modern affliction. It is a part of life in an oxygen-rich environment, and something with which all animals have to cope. Cancer is, essentially, cell proliferation, and cell proliferation is a part of life itself. It is essential for the development of life, for the development of the embryo, and for the renewal of the body of the adult.

This is one core reason why cancer is so difficult to treat. Cancer is not like a disease-inducing bacterium, or like a virus that infects the body. These pathogens are mostly easy to spot and relatively easy to differentiate from our own cells and tissues. Even if they mutate, which many pathogens do, we can generally still keep track of them and devise antibiotics to send out into the bloodstream to combat their growth. Cancer is different. Cancer cells are part of our own body, and this makes them difficult to mark as dangerous, and difficult to kill.

> Cancer is not like a disease-inducing bacterium or virus ... Cancer cells are part of our own body, and this makes them difficult to mark as dangerous, and difficult to kill.

There are three main methods of treatment in our arsenal at present: surgery, radiation, and chemotherapy. All of these are crude and often dangerous weapons.

Cancer has been known about since very ancient times. Malignant tumors have been found in the Egyptian mummies of around 5,000 years ago, and their effects are documented in writing. The Ebers papyrus, which emanates from ancient Egypt, is about 3,500 years old and there are descriptions of cancer in it.

It is possible that, in A.D. 180, Leonides of Alexandria performed the first surgical operation to treat a cancer. At least, his is one of the first known records of treatment. He removed the breast in women known to be suffering from breast cancer. Clearly, the severity and inevitability of cancer must have been well known for him to have performed such a major operation with the relatively crude materials and instruments he had at his disposal.

In essence, the principle behind surgical methods has not really changed that much since. Cutting the tumor out is perfectly possible as long as a cancer is not too advanced, and provided it hasn't spread to vital local tissues, or to distant organs. If the entire cancer colony is cut out, the treatment works, but sometimes the cancer is inaccessible to the

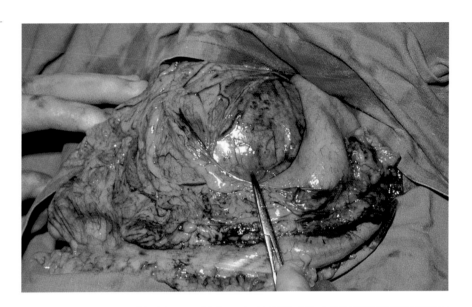

surgeon's knife, or just too large for this approach.

Much of the time, the mobile cancer cells may have already spread, and implanted secondary tumors in a different part of the body—most commonly, the lungs, liver, or brain, which have a particularly good blood supply. If the tumor has become "metastatic," or started to migrate between tissues and move around the body, surgery is sometimes impossible.

Radiation therapy and chemotherapy both control cancer by a different method. Ironically, both these forms of treatment damage the DNA. In doing so, they prevent cells from either dividing or replicating their DNA. Radiation therapy kills cells using carefully measured amounts of ionizing radiation directed as much as possible at only the tumorous tissues. Chemotherapy uses various combinations of drugs to achieve the same end. But because these drugs are given by mouth, or are injected into the bloodstream, they can have effects on healthy cells in the body, too.

There have been some marked successes with these therapies. For example, the mortality rates for leukemia and other childhood cancers have been considerably reduced, and chances of survival are now much increased for those with ovarian, testicular, throat, and lymphatic cancers and for some malignancies of the womb. A whole variety of treatments, which help to combat a considerable number of cancers, is now available; more and more cancers are cured by these methods and thousands of lives have been saved.

But radiation therapy and drug treatments are still rather blunt and occasionally quite cruel instruments. Cancer cells cannot be easily distinguished from normal cells, so both these methods can kill healthy as well as cancerous cells. They can also have severe side effects and a deleterious effect on general health. However, as the administration of these treatments has become more sophisticated, many of the most unpleasant side effects can be prevented. Massive hair loss, for example, or very pronounced nausea, are now much

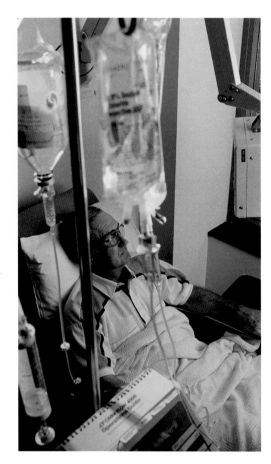

ABOVE *Chemotherapy is a cancer treatment that is improving hugely year by year.*
OPPOSITE *Top: A large cancer of the stomach. Bottom: Radiation therapy works by killing the most actively dividing cells.*

Lasers are used to line up the neutron beams before targeting brain tumors. This photograph is from Chicago, Illinois, but some of the first research into this technique was done at Hammersmith Hospital in London.

less common.

The irony of these methods is that they work by damaging DNA and are among the agents that, in other circumstances, cause cancer. In any event, they rarely kill all of the cancer cells. For this to happen, high doses would be needed and then there is the increased chance of normal tissue being damaged. But, most importantly, they can have a dangerous effect on the immune system. They suppress it, which can increase the chances of the cancer spreading, and immune suppression lays a patient open to infection.

But we should not blame the therapies for their limitations. Mel Greaves[1] persuasively argues the case for a Darwinian view of cancer. He describes how cancer cells become genetically unstable when the cancer is very advanced. They still produce clones, but DNA in the clones is not copied accurately, and so the new cells become genetically more diverse. This means that if we target the cancer with drugs aimed at particular kinds of cancer cells, there is a good chance that some of the mutated cells will not be affected. According to the laws of natural selection, these surviving cells will breed their own colonies of clones—clones that may be immune to the specific treatment being used. The difficulties of creating a well-targeted bullet to attack the cancer cells alone are immeasurably complex because we are dealing with a living, evolutionary successful colony that is hungry for territory and hungry for the body's resources.

We need a smart bomb, a magic bullet. Instead we have a bull in a china shop. So much for modern medicine. In the face of advanced metastatic cancers, we are often helpless. Hippocrates, in his *Aphorisms*, says, "it is better not to apply any treatment in cases of occult (literally "hidden") cancer; for, if treated, the patients die quickly, but if not treated they hold out for a long time." Perhaps, 2,000 years later, not much has changed; we still have to be careful that the cure is not worse than the disease.

1 Mel Greaves, *Cancer – The Evolutionary Legacy*, Oxford University Press (2000).

Iceland is a gift for geneticists—a living, thriving natural laboratory. The 275,000 people living on this cold and beautiful island are nearly all descended from a band of Norwegian settlers who landed on its shores more than 1,000 years ago. The original settlers had a hard time eking out a living in this barren place, and rapidly became isolated from the rest of the world; few if any other immigrants were willing to brave the harsh Icelandic conditions.

Thus, modern Icelanders carry stretches of DNA that have been passed down, unadulterated, since the "Age of Settlement." Iceland is, in the jargon of the moment, a closed system and ideal for studying the genetic factors that influence complex disorders such as cancer.

There are no family names in Iceland. Asgerdur Olafsdottir is so named because she is the daughter of Olaf; Kári Stefánsson is the son of Stefán. In order to ensure that the farms staked out by the original settlers were handed down to their rightful heirs, the people of Iceland kept careful genealogies. And so, in this land where few real trees grow, family trees have become a national obsession. No other people has such detailed records of its ancestry.

It is the Icelanders' well-documented family trees, coupled with their genetic isolation, that might allow scientists to track down the genetic roots of some kinds of cancer. Here, where almost everyone is descended from one small group of settlers, a phenomenon called the "founder effect" comes into play. Without outside influences to muddy the gene pool, a mutation found in the genes of one Icelander today can generally be examined in all of his or her relatives and can be followed back through the generations, to its source, or founder.

That traceable history helps researchers hone in on the gene that carries the mutation, and to work out if it is a factor in causing the cancer in question. So Iceland's genetic history has become an invaluable medical test bed.

Asgerdur Olafsdottir has never met Asa Jonsdottir—the two live on opposite sides of Reykjavik—but it doesn't surprise either of them to find out they are related. They are connected not only by a common ancestor, some five centuries back, but also by the genetic legacy they inherited from him, a legacy that has shaped both of their lives and the lives of their extended families. Both women are breast cancer survivors and both come from families riddled with the disease.

Asgerdur has a tragic tale to tell. "My grandmother—my mother's mother—got breast cancer and died from it. She had two daughters, my mother and her sister, and they both got breast cancer. My father had cancer as well—pancreatic cancer—and then my only brother had prostate cancer. Both my parents and my only brother died from cancers in a 20-month period. And, one year after my brother died, I got breast cancer myself. So it's quite a story."

Asa's family history is much the same story. And Asa herself was 40 when, in 1990, she first found a lump in her breast. She remembers remaining fairly calm. "I wasn't shocked," she says. "I said, 'Well, okay, this is the work that I have to go through.'" She had a mastectomy, followed by almost ten months of chemotherapy because there were signs that the tumor had begun to spread. Despite the medication, she found a second tumor in 1991, a tumor sitting in the scar left by the first surgery.

"I said, 'Oh, now I have to go to bed and put my blanket over my head,'" she says, somewhat self-mockingly. "Then I thought again, 'No, you can cope with this and everything will work out.'"

And it did. Ten years later, after a second bout of surgery and 24 more rounds of radiation, Asa is alive and healthy. She takes long daily walks with her husband and goes skiing in the winter. Asgerdur, five years out from her bout with the disease, is also doing well. Both women are mindful of the genetic legacy they have passed on to their children. Asa has three sons; Asgerdur has one. Obviously, they worry about their boys' health. Asgerdur has extracted a promise from her son that when he turns 40 he will get regular screening tests for prostate cancer.

Both Asa's and Asgerdur's families carry a mutated gene called BRCA2. This gene usually repairs DNA after damage caused by free radicals, or after cell division. When the gene is not working properly, cells are more likely to incur mutations that turn them into cancer cells.

> Our bodies are constantly producing cancer cells … but a healthy, vigilant immune system usually finds them and kills them before they have found a home and can begin to multiply.

BRCA2 is a vital factor in allowing breast cancer and prostate cancer to develop. Iceland has one of the highest incidences of breast cancer in the world; every Icelander who carries a mutated BRCA2 carries the same mutation, a mutation that prevents their damaged DNA from being repaired. Probably everyone who has breast cancer in Iceland on the basis of their BRCA2 gene is descended from a single individual, the so-called founder effect mentioned above.

This genetic warning light does not, at least at the moment, give us any more sophisticated techniques for treating these diseases. But there are clear benefits to knowing about BRCA2. People at risk can be offered genetic tests, or simply be alerted to keep a close watch on their health. Knowing they carry the mutation allows people to detect the disease before it has done too much damage, and before it has started to spread.

Asgerdur credits this vigilance with saving her life; her cancer was found on a mammogram after she discovered a lump during a routine self-examination. Unfortunately, breast cancer is the exception, in having a known genetic marker, rather than the rule. There are still very few known inherited cancers, where it is easy to pinpoint a particular genetic marker.

This is where Kári Stefánsson and his company, deCODE Genetics, are really hoping to shine. The scientists at deCODE have already begun combing through the genome of Icelandic patients who have had lung cancer. Although, of course, it is very well known that smoking can lead to malignancies in the lungs, there are a number of unanswered questions. Why do some smokers develop cancer while others do not? Could it be that some people have a genetic predisposition toward lung cancer? Early research suggests this could be the case. And if there is any place that they might be able to track down that predisposition, it is in Iceland, where the gene pool is still limited and where there are excellent chances of its proper documentation.

To find any genetic component of lung cancer will require larger and more complex studies; huge numbers of people will need to be carefully screened for suspicious bits of DNA and then monitored closely over many years. DeCODE is very well poised to do these studies. The company has been granted permission by the Icelandic government to collect the genealogies of the Icelanders, as well as their individual medical records.

This database will be cross-referenced with the results of blood tests performed on patients at apparently high risk for any number of conditions. Once that is accomplished, says Stefánsson, deCODE may be able to untangle the genetic roots of many forms of cancer, lung cancer among them, as well as any number of other common diseases. Coupled with the information streaming out of the various laboratories decoding the human genome, this database, he claims, will be a veritable treasure trove of information.

Constructing a database of an entire nation

Top: Mammography is a powerful screening tool in breast cancer. The calcified white area at the top of the image is a cancer. Above: Self-examination may be helpful in getting an early diagnosis of this increasingly common disease.

is, to say the least, highly controversial. Moreover, commercialization of such a project is also a major bone of contention. But even for a small country, establishing a database of this size is incredibly expensive, costing tens of millions of dollars. DeCODE will fund the mammoth task of collating all the available records and will provide the tests and their results free to Icelanders. The Icelanders themselves have the right to opt out of the database altogether and between 10 and 15 percent have done just that.

In return for giving the results of any test free to any member of the Icelandic populace, deCODE has negotiated the right to sell the information it gleans from the Icelandic genome to international pharmaceutical companies. These companies, of course, hope that they may then be able to devise better therapies, tailored to the individual with a known specific genetic predisposition.

It is no surprise that Stefánsson and his company have their critics. Many people in Iceland—and around the world—are worried about putting such sensitive information, normally shared only between doctor and patient, in the hands of a private company. They worry that the data may make its way to insurance firms or employers, and that it may be used against individuals found to carry genes for cancer or other debilitating—and costly—diseases.

The breast cancer genes (BRCA1 and BRCA2) are a perfect example of these concerns. Women who apply for life insurance could possibly be told that they will have to have a test that shows whether they have a mutation on either one or both of these genes. And if they have the mutation, they could be refused insurance or the premiums they pay may be very high.

Stefánsson acknowledges the ethical concerns, but defends his program. "The question this leads to is whether it would have been better not to have made this discovery of breast cancer genes because it can be abused. My answer to that is that it would have been a crime to suppress this knowledge because it can be used to save lives. I think it's important to recognize that we should not let crooks write the rules. We should pass laws that forbid the abuse of genetic information."

The BRCA1 and BRCA2 mutations are not necessarily fatal, but they make cancers much more likely. The accumulated damage to DNA over time means that a cell's mutation into immortality, the mark of a cancer cell, is more likely to happen. This is why cancer becomes increasingly more pervasive as we grow old. Kári Stefánsson believes we must accept that cancer rates will rise in the immediate future. He says, with a touch of irony, "Early death is the best prevention for diseases with late onset. That's not a surprise, and that's not specific to cancer. We should expect that cancer is going to kill more and more people in our society, because the population is becoming older." There is no escaping demography.

Heart disease and Alzheimer's disease have also been rising in incidence—a result not only of lifestyle changes, but also of our ability to beat back many infectious diseases with antibiotics or antivirals, to vaccinate against most lethal childhood diseases, and to bear children with a much greater degree of safety. The longer we live, the more likely we are to encounter—over and over again—the various environmental toxins that add to the already prodigious free radical hordes nibbling away at our DNA. Each pollutant-laden breath we take is an assault on our lung cells. Smoking makes the insult several times greater. Each time we step out into the sun, the ultraviolet rays are given another chance to destroy our skin cells. The cells may be able to bounce back from the initial assault—or the tenth, the hundredth, or thousandth—but the chances become greater and greater that they will concede defeat.

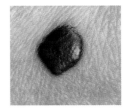

When the situation is viewed in this slightly depressing light, it seems amazing that any of us avoid this affliction. But most of us are well engineered to deal with the threat. Our bodies are con-

This malignant melanoma on the forearm grew very rapidly. When it was photographed, it was about the size of a dime.

stantly producing cancer cells. Mutant clones are commonplace in our tissues and our bloodstream, but a healthy, vigilant immune system usually finds them and kills them before they have found a home and before they can begin to multiply.

This means that when the immune system is defective, suppressed, or malfunctioning, more mutant cells will slip through the net. Malignant melanoma is one example of a cancer in which the immune system is thought to be a key factor. It was thought that ultraviolet (UV) light from the sun bombards our cells and directly damages the DNA, causing more mutation than is normal, and increasing the risk of cancer cells turning into melanoma. In fact, new theories suggest that sudden doses of UV light suppress our immune system. This may be why we often come back from sunny climates with a cold. The more serious result is that the white cells patrolling the skin are few on the ground, hence stray melanoma cells can go undetected and multiply more readily.

It also suggests that, potentially, the immune system could be artificially kick-started into attacking more advanced cancers. This idea has given rise to an extremely powerful and potentially rewarding area of research—immunotherapy and the search for a cancer vaccine.

There have been three attested "miracles" at Lourdes involving cancer since the Second World War. All of them involved advanced cancers that quickly, and spontaneously, went into remission. One of the recipients of these miracles was an Italian army officer in his 60s, whose hip had practically been destroyed by a sarcoma of the pelvis. He had to wear a full body cast, and was unable to walk. In 1976 he made the trip to Lourdes and as he was immersed in the water he claimed to have felt an electrical charge flowing through his body. Immediately he regained his appetite, which he had lost after gangrene had set into his leg. Doctors at the International Medical Commission at Lourdes, who investigate such cases, took off the cast and X-rayed his hip. The tumor appeared to have shrunk. Over the following weeks the hip joint reformed and within two months he was walking.

According to the Roman Catholic Church this case passed the necessary tests and checks to qualify as a genuine miracle[2], but there is also a possible medical explanation. A number of recorded cases of spontaneous remission have occurred when a cancer patient has contracted a viral or bacterial infection. The theory goes that an infection can provoke a strong response in the patient's immune system, a response that can not only fight the infection, but also galvanize a powerful counterattack on the colony of cancer cells. For some reason the infection reminds the immune system of the need to fight the tumor. In fact, there have been many cases of cancers that spontaneously remit, particularly skin cancer, and many people think that infection is the chief cause.

The idea of using infection as a cancer vaccine has been around for some time. In 1938, William B. Coley, an eminent New York City surgeon, stumbled across a phenomenon that was to change the course of his life's work and make a promising contribution to the fight against cancer.

Coley became frustrated after losing a 19-year-old patient to bone cancer, despite early detection, amputation of her arm, and a relatively good prognosis in that particular patient. So he trawled through records at this New York hospital of every bone cancer patient; most cases had ended in death. But there was one survivor who made a remarkable recovery; the man's doctors had given up, but one day the cancer had

---

2 Carl Sagan (Lynn Margulis's first husband) suggested that, statistically, Lourdes has not had its fair share of spontaneous remissions. Given that the spontaneous remission rate for all cancers is between one in 10,000 to one in 100,000, there should have been something between 50 and 500 "miraculous" cures of cancer alone. He says "the rate of spontaneous remission at Lourdes seems to be lower than if the victims had just stayed at home." Carl Sagan, *The Demon-Haunted World*, p. 232.

gone into remission and the patient walked out, seemingly in perfect health. Coley examined the records. Just before his recovery, the man had suffered two attacks of erysipelas, a severe skin infection caused by the bacteria *Streptococcus pyogenes*. Coley wondered, could this infection have somehow affected the cancer?

He injected other patients suffering from advanced cancers with *streptococcus* cultures, but he had no success until he managed to find a particularly virulent strain. A male patient with tumors on his tonsils and neck was injected; he developed severe erysipelas and a high fever, but, within a few days, the tumors were gone. The infection, it seemed, had awakened his immune system, which had suddenly, and effectively, attacked the cancerous cells.

Coley published his results and set out about refining the technique. Treatment with the live bacteria was too dangerous, so he mixed the strep-tococcal culture with another bacterium, *Serratia marcesens*. He claimed some success in treating inoperable cancers, but the unpredictable outcome of his approach meant he was ignored by the med-ical establishment.

Coley's toxins, as the treatment was called, made sporadic appearances in the medical journals from the 1940s until the present day. But no one took it seriously. The success of the toxins was constantly in dispute, and over the years regular accusations have been made that "the establish-ment" repressed the results of Coley's work in favor of more traditional techniques such as chemotherapy. Coley never made it to the main-stream of cancer research, and never gained recog-nition. The success of his methods remains something of a mystery.

Nonetheless, the idea of a cancer vaccine gained credibility, and, in 1971, it became a major part of Nixon's war on cancer. But there have been no great breakthroughs. Only in the past few years has our knowledge of the intricacies of immunology given us the means to refine the technique.

Jeff Allan, literally a miraculous survivor after experimental cancer treatment. His kind of story will become an increasingly common one with progress in cancer treatments.

Jeff Allan's wife, Dawn, first spotted the mole on his back while the two were on vacation in Cornwall, England, in 1993. Every year millions of fair-skinned people roast their white bodies under the sun, the most potent powerful carcinogen in the solar system, and thousands suffer the consequences. Jeff's mole turned out to be a malignant melanoma, a potentially lethal form of skin cancer.

To his relief, the doctors said they had diagnosed the cancer early and the tumor had not yet burrowed too deeply into his skin. He might be lucky. After the mole had been removed, his doctors were optimistic about Jeff's chances. However, this particular cancer is a devious disease. Melanomas constantly shed cells, which are picked up in the bloodstream or the lymphatic system. These daughter cells travel around, allowing the disease to spread deep inside us unseen and unfelt.

Jeff went back to be checked regularly—every month or two—over the next year. Then, feeling as if they had somehow dodged the bullet, he and Dawn decided they would have a second child. Dawn got pregnant immediately, but then, just a few weeks later, Jeff felt a lump under his right arm.

It was, once again, a melanoma. Jeff underwent another operation to remove the new tumor, but it returned almost immediately. They operated again. They blasted Jeff with radiation therapy, only to discover— just after their son, Luke, was born—that the peripatetic cancer cells had invaded his spine. Jeff went back to have yet more radiation therapy. He remembers wondering if he would get to see his son grow up. This was, without a doubt, the worst of times.

But he didn't suffer self-pity. He had a new son, a four-year-old daughter, and a wife who needed him to stay alive and fight. Jeff heard about a project that was exploring the possibilities of a melanoma vaccine, headed by virologist Angus Dalgleish from St. George's Hospital, in London. Immediately, Jeff knew he had to be referred to Dalgleish; that he couldn't take no for an answer. Initially, no was the answer he got. But Dr. Dalgleish was just about to start a second trial using a species of bug from the mycobacterium family. It was not long before Dr. Dalgleish suggested Jeff take part.

The mycobacteria are a large bacterial group, containing the microbes that cause tuberculosis and leprosy. But the mycobacterium in Dr. Dalgleish's vaccine—*Mycobacterium vaccae*—is not a pathogen, or disease-causing agent, but a soil-dwelling microbe. As Dr. Dalgleish points out it is still a foreign body, and, when injected into the human body, it wakes up the immune system. In a cancer patient like Jeff, Dr. Dalgleish says, the hope is that it will give the immune system a nonspecific boost to persuade it to hunt down not only the bacterial invaders, but the errant cancer cells as well.

This hope turned out to be well founded. Dr. Dalgleish had already conducted an early phase trial of the vaccine in ten patients with advanced, difficult-to-treat prostate cancer. Of those patients five showed a decrease in their prostate specific antigen, or PSA, levels. PSA is the standard marker of prostate cancer; the higher the number, the more advanced the cancer. Two of those patients who had responded had received no treatment other than the vaccine; in other words, there was no other explanation for their improvement. A small sample, but tantalizingly successful.

Given that he would surely die without some miracle, Jeff began traveling to London every two weeks at first, then monthly, to receive his injections. Initially, the blood tests did not show much of an immune response. But after half a dozen injections Dr. Dalgleish saw some progress. The immune cells were beginning to regroup in Jeff's bloodstream.

Dr. Dalgleish and his colleagues reported that, of the 24 patients who received the vaccine in these early trials—trials that are designed to test the vaccine's safety as well as its effectiveness—almost 40 percent showed an immune response. And those patients who did respond survived, on average, 28 weeks longer than those who did not. Seven months is not a terribly long time, but these patients' cancers were at an advanced stage, and, most importantly, the principle seemed to be working.

Jeff responded spectacularly well. There was a huge surge in his immune system and the cancer started to retreat. In fact, everything began to look so promising that at one point the doctors questioned whether or not Jeff had really been as ill as he had seemed. Perhaps, they thought, his records had been mixed up with those of another patient. Dr. Dalgleish even wrote to Jeff's original doctors in Sheffield to be sure his disease had been as bad as reported.

Today, almost seven years after his initial diagnosis, Jeff still has his injections every three months, and tests continue to show the cancer is in remission. He says, happily, "The bones are re-forming, the scans look good, the CTs look good. Everything looks good, and I'm absolutely delighted with it."

Like all patients who have lived with advanced cancers, Jeff cannot rest on his laurels. He must continue with the vaccine injections and be checked regularly for any potentially threatening lumps. "I don't know if I have been cured. I know where I was yesterday, I know where I am now, I don't know where I'm going to be tomorrow. No one will turn around and say, 'you're cured.' I wish they would, but I don't think they'll ever say that because they don't know." Nevertheless, compared to his condition just after Luke was born, these are the best of times.

The cancer vaccine, so long a dream for people like William Coley, may have come of age. This is true Superhuman medicine; it relies heavily on the body's own ability to seek out intruders. Vaccines will not be a miracle cure; they will almost certainly not be as successful a magic bullet as, say, penicillin. However, they may become part of a wider system of managing cancer as a long-term chronic disease.

We simply do not know how to banish every rogue cell from the body, especially when dealing with advanced metastatic cancers. The evolutionary cunning displayed by these genetically volatile mutant cells, and the pressures of selection that allow for the survival of the colonies when we try to kill them with radiation or chemicals, mean that we are constantly trying to catch up. Still, perhaps we can combine various methods of therapy and keep the cancer dormant. Cancer may, in the future, come to be viewed in the same way as diabetes—an incurable condition but one that we can keep under control so that the patient can live a relatively normal life.

A few kinds of cancer have begun to crack under the weight of the enormous scientific research effort over the past 30 years, but mesothelioma is not among them. It is one of the most malignant and fast-moving cancers around. In its advanced stage, mesothelioma is considered incurable. If the cancer is found early, and treated aggressively, at best only some 20 percent of patients will survive. In most mesothelioma patients, however, the cancer grows silently and for such a long time that even that unimpressive statistic is beyond their reach. When it affects the lining of the lung cavity, the most common form, the disease leaves sufferers increasingly short of breath, weak, and prone to infections. Chemotherapy and surgery only provide temporary relief. The average survival time of an advanced mesothelioma patient is one year.

In the UK, more people die from asbestos-related disease than in traffic accidents.

Mesothelioma is a cancer of the cells that line either the chest cavity (the pleura), the abdominal cavity (the peritoneum), or the cavity around the heart (the pericardium). About 75 percent of mesotheliomas are pleural in origin; most of the remainder are peritoneal. Its cause, in the majority of cases, can be summed up in a single, chilling word—asbestos.

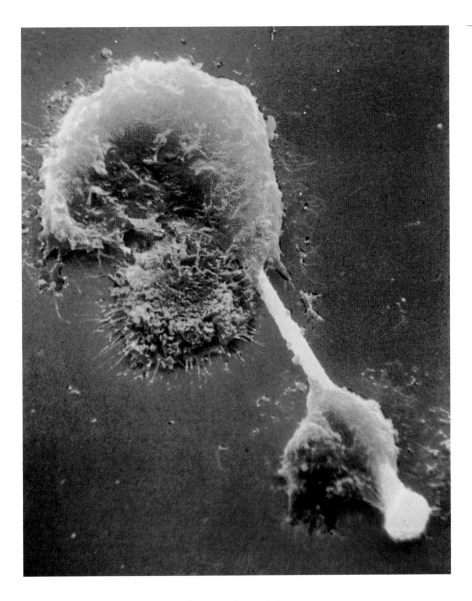

After working with asbestos for a lifetime, the chances of contracting mesothelioma are frighteningly high. However, because these natu-

An asbestos fiber being partly engulfed by two white cells. The asbestos is winning the battle and it will produce one of the most virulent cancers known to man.

ral silicate fibers persist in the lungs, unlike tobacco smoke, the risk even from transient or intermittent exposure is very disturbing. In the U.K., more people die from asbestos-related disease than in traffic accidents.

In the United States, about 2,000–3,000 people are diagnosed with mesothelioma each year. However, according to the American Cancer Society, the number is increasing, and men are three to five times more likely than women to contract the disease.

Because the disease takes so long to develop, the incidence of mesothelioma is expected to triple over the next 20 years in the UK. One percent of men who were born in the 1940s will die from mesothelioma. There were about 1,300 deaths from mesothelioma in 1996, which was greater than the number of deaths from cervical cancer; the difference is that public awareness of mesothelioma is virtually nonexistent. Mesothelioma is really a silent epidemic. There are 100,000 men in the UK, who, having been exposed to asbestos, are currently free from symptoms, but they will die from mesothelioma during the next 30 years.

In 1974, Bryan Newton became a joiner on an oil rig. His job was to build accommodation modules out of a material called Maronite that was impregnated with asbestos. He was not aware that he should be taking any precautions and the management did not inform him. Eventually, after a year, a doctor came down and tested the air; it contained more than six times the recommended concentration of asbestos. By then Bryan had spent a year breathing in the dust and it was too late.

In 1997 Bryan began to feel short of breath, which he put down to a problem with his heart—he had had a heart attack ten years before. Then, a year later, the 53-year-old felt severe chest pain. He was rushed to the hospital, where a cardiologist said there was nothing wrong with his heart, but he had found some fluid in Bryan's chest. By January his consultant had enough information to give Bryan a diagnosis that was little more than a death sentence: advanced malignant mesothelioma. Bryan asked what treatments were available, to which the consultant replied, "None." Bryan was told he had between a few months and two years to live.

Bryan could not sit back and wait to die. He began to do his own research, and eventually found himself flying to Pennsylvania, to meet Daniel Sterman, director of a a gene therapy trial aimed at obliterating pleural mesothelioma. It was a trial, Bryan hoped, that might just save his life.

Sterman says that Bryan was typical of the patients he often sees, patients who may know more about the disease or other therapies than he does. He finds them very well informed, very aware of the various options, and willing to travel; most importantly, they do not accept the treatments that their doctors back home have offered them. They are searching for more effective alternatives. Sterman needs patients like Bryan. He needs new patients to try out the new drugs and new protocols. It is important for the researchers to get the patients, and it is also important for the patients, who can have a chance to participate in experimental therapy, which at least has a theoretical chance of saving their lives.

Bryan's goals were clear. He was not willing to accept the death sentence his doctors had given him, nor was he willing to allow them to destroy whatever time he might have left. He wanted to maintain a good quality of life, not just be propped up by drugs.

His doctors had talked to him about some chemotherapy trials, but Bryan was loath to suffer the drugs' side effects for what most likely would only amount to a short extension of his life. And then, of course, there was the option of surgery.

Surgery is the venerable old workhorse of cancer treatments, and the philosophy behind it is certainly simple enough: Go in, cut out the tumor, sew up the incision, and let the patient get on with life. But the cancer cells may be too dispersed, may already have moved into the bloodstream to colonize other sites, or they may never have been organized into a solid tumor to begin with. Bryan's mesothelioma was certainly one cancer that was beyond the reach of surgery. An operation would provide him with relief, but it would not cure him. And so he wanted to keep it only as a final option.

Gene therapy, on the other hand, is a truly Superhuman procedure, a therapy that uses the cell's mechanisms to cause it to self-destruct. Gene therapy is a cutting-edge treatment that offers great hope—if not now, at least in the future.

The experiment Dr. Sterman and his colleagues were conducting involved a crippled version of an adenovirus, a virus that has an affinity for cells in the respiratory tract, which carried a so-called "suicide gene" that would be delivered to the tumor cells. The virus is modified in such a way that it will infect only the rapidly dividing tumor cells, leaving the rest of the lung and chest wall alone—at least, that is the theory. Once having been infused into the patient's chest, it should prompt the cells of the tumor to pump out an enzyme called thymidine kinase, making the cells appear as if they were virally infected. That would be followed by an intravenous assault with an antiviral drug, which kills only cells that are producing thymidine kinase.

Bryan thought the whole procedure sounded eminently logical. Dr. Sterman's gene therapy trial might very well have offered Bryan a chance, but Dr. Sterman and his team have a long waiting list. Their biggest limitation has been the quantity of the gene therapy virus available. Like all synthetic viruses used for gene therapy, it is very complicated to produce.

Moreover, a huge amount is needed, enough to treat each patient for nine months.

Bryan's tumor was not willing to wait. In just the few weeks after Dr. Sterman had received Bryan's medical reports, fluid had accumulated in Bryan's chest, and the tumor itself was rapidly growing again. Dr. Sterman put Bryan on the list, but, even so, his treatment would not start for several months; by that time his chest cavity could be obliterated with the tumor, leaving no space to infuse the gene therapy. To try to buy him some time, Bryan started a course of chemotherapy but the drugs were unable to stem the growth of the cancer. Bryan died soon after.

Trials of this kind are fraught with ethical dilemmas, and scientists such as Dr. Sterman are careful not to offer false hope. Inevitably, new treatments can be tested only on a small number of people; there will always be disappointments, and sometimes tragedies.

Dr. Sterman has some success stories, too. Donald Hardy, 68 years old, and his wife, Joan, are living out their retirement in the warmth of Florida, after Don had beaten the odds to overcome a diagnosis of mesothelioma, which seemed likely to kill him within months. Don had worked with asbestos for 35 years. Like Bryan, he was told there was no conventional treatment, but, having done some research, he became the very first patient to take part in Dr. Sterman's gene therapy trial.

No one can say for sure how much of a role the gene therapy played in Don's survival. There are occasional cases of mesothelioma going into remission, but the fact remains that it is extremely unusual for a mesothelioma patient to survive for five years.

By the end of 1999, Dr. Sterman had completed the second phase of his trial, and was ready to analyze the data. Then Jesse Gelsinger died (see pages 113 and 248), the first gene therapy patient ever whose death was attributable to the use of viruses to carry the gene. That trial was taking place at the University of Pennsylvania, and Gelsinger's death threw the entire Pennsylvania gene therapy program into disarray. The next phase of Dr. Sterman's trial was put on hold indefinitely. Sterman is philosophical. "When people hear about gene therapy, they expect that, within a year or two, it should be available as easily as treatment for the common cold."

OPPOSITE A natural killer T cell (yellow) attacking a cancer cell (orange). This property of the immune system is one of the most important defenses against cancer.

He realizes that people both inside and outside the profession need to be patient. "It took many, many years for antibiotics to be developed, and I think it's going to take many years for gene therapy to be perfected. Are we as a society going

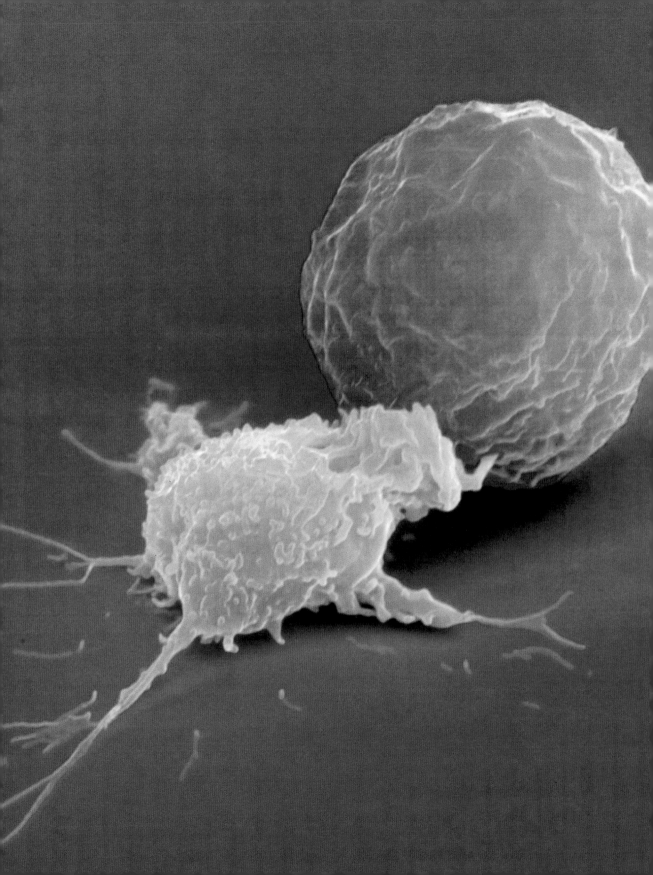

to have the patience to allow the researchers, scientists, and clinicians to work out the kinks and make this an effective therapy? It may be 20 or 30 years. I am willing to put in the time, but the question is, will the funding be there for the trials and the research? Or, will the public become disenchanted that it doesn't seem to be working and withdraw the funds?"

Gene therapy is still experimental, and long away from the mainstream of cancer treatment, but many researchers in the field are convinced of its potential. They hope that time and money will pay off. Dr. Sterman emphasizes that gene therapy is still in its infancy: "I liken it to the development of the airplane. Where we are right now is at the Wright brothers' Kitty Hawk stage, with a rickety airplane made out of kite parts and an old automobile engine, flying across the sand dunes, trying to see if we can get up into the air for more than 20 seconds. We're a long way from having a Boeing 747. But if we have enough time and resources, we will get to the point where we have that 747—where gene therapy, genetic medicine, will be routine."

We are dealing with a slippery and mercurial affliction. Treating cancer is a little like killing weeds. We can dose the invading plants with high-strength herbicides, which might kill the majority of weeds, but the chances are there will be one or two strains hidden away that are genetically resistant to the poison. These strains will multiply even more readily, now that they have more space and more nutrients to flourish. They will spread, and our herbicide will be useless against them. The herbicide applies selective pressure, and the evolutionary process ensures that the weeds will be back, fitter and stronger than before.

Chemotherapy and radiation therapy also apply selective pressure, encouraging the emergence of resistant strains of cancer cells. As Mel Greaves points out, this is an irreconcilable conflict for many cancer therapies, especially once the colonies have grown more advanced, and are, therefore, genetically more diverse; the therapies themselves cause stronger mutant cells to come forward and multiply.

Maybe there is a way to avoid this paradox. Once again the approach lies in taking advantage of our own biological systems, and knowing how to intervene to best effect.

Angiogenesis is the process by which tumors set up and exploit a blood supply to keep themselves alive. No cancer cells, whatever their genetic makeup or resistance to chemotherapy, can survive without blood, so to cut off their food and oxygen source is an interesting way to go about killing the cancer. It should not apply any selective pressure, because, in this instance, there is no possibility of mutated cells living without blood. And the cells making up the blood vessels themselves are not cancer cells—they are genetically stable—and so would not produce new, resistant strains to form new vessels.

The search was on for a drug that could block the cancer's lifeline. The search for new drugs, however, is one of trial and error, with many false leads and dead ends. What looks promising in the laboratory may not work in human trials, and many exciting leads prove to be disappointing. But anticancer drugs can crop up in the most unexpected places. For centuries we have known that yew trees are poisonous. Just a handful of yew leaves could make you extremely ill; yew leaves, however, were the original source for two highly effective anticancer drugs, taxotere and taxol.

> … anti cancer drugs can crop up in the most unexpected places. For centuries we have known that yew trees are poisonous … yew leaves, however, were the original source for two highly effective anti-cancer drugs …

Pharmaceutical companies have been scouring the world for naturally occurring drugs that might have the effect of cutting off the blood supply and therefore suffocating the tumor. They may have found such a drug, a substance derived from the bark of the African bush willow tree. Zulu witch doctors used it for centuries, but, in their case, as a poison as well as a medicine. They applied it to the tips of their arrows to poison their enemies.

Combrestatin is a synthetic mimic of this natural pharmaceutical, and it is being tested against a number of different, highly vascularized cancers at Mount Vernon and Hammersmith Hospitals in London. In 17 patients, this tumor-starving therapy was able to cut off the blood supply to breast, bowel, and lung malignancies.

Two years ago, headmistress Diane Ransom went into the hospital for a routine hysterectomy, and the surgeon found a tumor in her womb. Then doctors discovered it had spread to the glands in her groin. A second tumor, now the size of an orange, was growing so fast it was starting to squeeze out healthy tissue. Diane had two rounds of chemotherapy, which did not work. If there was no effective method of killing off the cancer directly, it might be possible to set up a blockade and starve it to death? Diane decided to take part in the Hammersmith Hospital trial. Maybe an ancient Zulu potion would work where sophisticated chemotherapy had not.

The process of finding new drugs for cancer treatment is full of false turns and blind alleys. Here I am in the maze at Longleat House, surrounded by yew leaves, which produce one of the most powerful anti-cancer drugs.

The results show the toughness of some cancers. Diane was injected with the drug. She quickly began feeling pain in her abdomen, a good sign; it meant the vessels were constricting and the blood supply to the tumor was cut down. Scans confirmed this. Then there was nothing to do but wait and see.

The doctors were optimistic, but after several weeks of injections, the tumor had not shrunk. The blood supply had been cut off almost entirely, but only for a while; the tumor was hardier than they thought. The drug was also proving to be painful and debilitating, and Diane had to stop the trial.

It is no great comfort to Diane, but the fact that the blood supply to the tumor dried up, even temporarily, demonstrates "proof of principle" experimenters are on the right track. The next step is to try higher doses, or to use the drug in combination with other therapies. The trial researchers emphasize that antiangiogenics will not be a complete cure; they won't be capable of killing every malignant cell in the body.

Similar trials are taking place, using a variety of strange substances, including compounds derived from shark's cartilage, green tea, and, incredibly, thalidomide. It would be ironic if this infamous drug, used in the 1960s as an apparently harmless sedative, turned out, after all, to be a lifesaver. Thalidomide was the drug that caused so many children to be born with limb deformities after their mothers took it in early pregnancy. It caused these deformities by interfering with the growth of blood vessels during early development. Almost more than any other event, the birth of those babies demonstrated that medical researchers were far from infallible. It was this major accident, more than any other, that first raised public suspicion about the dangers of modern medicine and medical overconfidence. But thalidomide is safe in people who are not pregnant. Given its special potential, there is quiet optimism that it and other antiangiogenics, particularly if used in conjunction with other therapy, will turn out to be very useful in managing malignant disease.

> … we should be looking for ways to control the disease, to live with it, and to thrive despite it.

Nixon's all-out cure for cancer may be a dream that, realistically, is unattainable. Managing cancer may be as much as we can hope for, so we should be looking for ways to control the disease, to live with it, and to thrive despite it. Asthma is one such disease that we can now, by and large, control. Fifteen years ago AIDS was a death sentence; now, with current treatment, people can live five, ten, sometimes fifteen years with HIV, perhaps even longer. During that time span it is a chronic but manageable disease. Cancers are much the same. Many cancers will be fatal, but many can be managed and kept under control.

Kári Stefánsson believes that we have to set ourselves realistic goals. We are not going to get rid of cancer in one fell swoop. "There will not be a glorious moment when a man, standing on the moon, waves to people down on earth saying, 'We have gotten rid of cancer once and for all!' That's not going to happen."

Cancer will probably be around as long as our species exists. The issue is our ability to tame it, to control it, and ultimately to live with it.

5 Outbreak

Five hundred years before Christ, ancient Athens was alive with the creative spirit and social vitality that led to the foundations of Western civilization. But in the year 430 BC, that cradle of modern democracy was fatally attacked. The war with the Spartans forced the Athenians to crowd within the city's walls. The deprivation, hunger, and lack of a clean water supply during that siege left them sitting ducks for infection. One type of microbe in particular seems to have prevailed. First came the fever, then headaches, the chest pain, the projectile vomiting, the unquenchable thirst, the uncontrollable diarrhea, dehydration, and, finally, total collapse.

Thucydides, the famous writer and upper-class Athenian who survived the war, wrote in his *Histories*:

> *As a rule, however, there was no ostensible cause; but people in good health were all of a sudden attacked by violent heats in the head, and redness and inflammation in the eyes, the inward parts, such as the throat or tongue, becoming bloody and emitting an unnatural and fetid breath. These symptoms were followed by sneezing and hoarseness, after which the pain soon reached the chest, and produced a hard cough. When it fixed in the stomach, it upset it; and discharges of bile of every kind named by physicians followed, producing violent spasms. Externally, the body was not very hot to the touch, nor pale in its appearance, but reddish, livid, and breaking out into small pustules and ulcers. But internally it burned so that the patient could not bear to have on him clothing or linen even of the very lightest description; or indeed to be otherwise than stark naked.*

Then, of course, came death. There were so many deaths and the people of the city were so debilitated that corpses lay unburied in the street. It is said that so many corpses turned so rotten that even the birds of prey refused to peck at the offerings. Within three years at least one-quarter of the population of Athens was dead. To this day the precise cause of the illness remains a mystery, the various descriptions of it do not quite fit cholera or bubonic plague. Nonetheless, there is little doubt that some form of bacterial infection was the culprit.

To its citizens the Great Plague of Athens came down from on high. It descended with such fury that it could only be understood as punishment by the gods for the sins of the mortals. Today, we can look at it for what it really was: a horrific example of one of the most natural disasters.

In many ways the devastation wrought by volcanoes, earthquakes, hurricanes, and floods alone is

The Great Plague of Athens *by Michael Sweerts* *(1624–64).*

never quite as profound. Even today, it is the human infections that so frequently follow such catastrophes that still cause the most devastation. Bacteria are the most primitive of life-forms, the simplest, and among the smallest. They each weigh around .00000000001 gram, yet once a few start to reproduce, they can sicken or kill a blue whale weighing 100 tons. And viruses are even smaller than bacteria, and apparently more insignificant.

We know now that the spread of epidemics is as natural as the movement of planets in the sky. Most of us do not believe disease is a punishment or a portent. We do not fear that a solar eclipse is a sign of an impending plague. We know now that our bodies have been engaged in an endless arms race with these microscopic life forms over millions of years, and our lives—and deaths—have been shaped by this perpetual battle.

A few decades ago, it was a race that everyone thought we were winning. Penicillin, the polio vaccine, antibiotics, and massive public health drives all combined to reinforce this impression. Doctors enthusiastically set out to immunize the world and target bacterial infection using antibiotics, magic bullets that attacked these diseases cheaply and effectively.

The effect on public health was immense. One hundred years ago, two-thirds of children died before adolescence. A simple cut could lead to a lethal blood infection such as septicemia. Polio could kill or cripple a child for life. Once vaccines and antibiotics became commonplace, life expectancy increased by decades. They are without doubt the great life-savers of modern medicine. In 1967, the US Surgeon General declared that infectious diseases had been conquered. It was only a matter of time, he said, before every plague that had struck fear into our hearts would be a distant memory. His optimism was catching. Everyone started to believe we were on the way to eradicating infectious disease from the planet.

The World Health Organization actually set up a timetable for wiping out infectious diseases. The first on the list was smallpox; and WHO duly checked it off in 1977, when the disease was finally eradicated. By the year 2001, there was to be no more tuberculosis. Other diseases were scheduled for their demise—measles, polio, and diphtheria. There was nothing, it seemed, we could not kill or, at the very least, control.

But decades of success led to complacency and the pace of research into bacterial and viral infection slowed. This was a mistake; we simply did not anticipate what would happen next.

New and lethal viruses began to appear. No one knew where they came from and how they emerged. Some were entirely different from all known viruses, and many caused horrific diseases. They included the Hantaan, Junin, Rift Valley Fever, Lassa, and Ebola viruses.

There is an isolated cave on the border of Uganda and Kenya that is thought to be where the Marburg virus originated. Marburg, or Green Monkey Disease, is a virus similar to Ebola. Sufferers start with a fever and end up with both internal and external bleeding. Death is painful and inevitable. Two people contracted Marburg after visiting this cave, and both died within a few days.

In 1988, a group of Ameican and Kenyan scientists visited the cave to search for this deadly virus. They found nothing. Had the virus come and then departed? Or was it coincidence that the two visitors had contracted the same unknown disease? How long had the virus lain dormant? And how many other lethal viruses are hiding in nooks

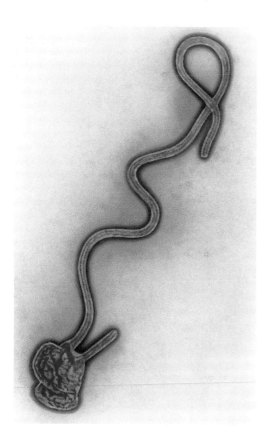

ABOVE The Ebola virus. You start with a bit of a fever, but hemorrhage and death are almost inevitable. Sometimes the beauty of the organism belies its viciousness.
OPPOSITE Penicillin mold growing in a culture dish. It is extraordinary how this nondescript organism has saved so many lives in the fight against bacteria.

and crannies, waiting for their chance to infect the population? No one knows. But it is likely that the disease will reemerge at some time in the future. And it is a terrifying thought that there is a vast untapped reservoir of bacteria and viruses lying dormant in the world around us.

There is a danger that eclipses even these new and exotic diseases, a danger which, to a large extent, is of our own making. Over the past 50 years medical practitioners have thoroughly embraced antibiotics. We use them liberally, even for minor infections such as a sore throat. They are a cheap and easy form of infection control—better to take a few pills than months convalescing in the Swiss Alps. But there is a price to pay for what many people consider to be the overenthusiastic use of antibiotics. Old infections are coming back to haunt us, in the form of new and powerful superbugs.

Tim Streatfield was sitting in his usual spot in a first-class carriage of a train one October morning. The Great Western express from Cheltenham barreled down the track toward London and, at 8:11 AM, the train began its approach into Paddington station. Tim had no idea that just ahead a Thames commuter train had left the station, passed a red signal, and was traveling directly into the path of the Great Western train at Ladbroke Grove.

It was Britain's worst rail accident in more than a decade, killing 31 people and injuring over 200. Tim was fortunate to be part of the latter statistic. Almost half of his body was burned when his carriage burst into flames. Until quite recently, burns of this severity would usually mean certain death. With better ways of dealing with collapse, dehydration, and skin loss, modern medicine is far more successful at saving the lives of such victims of fire. But Tim's greatest medical challenge lay ahead in the form of a microbe that, in a way, is a child of modern medicine—a microbe that has the ability to evolve into a more robust life-form. The story of Tim Streatfield is an increasingly common cautionary tale of what is becoming known as the postantibiotic era.

Treating a patient in an intensive care unit is like holding a bacterial open house. Catheters, intravenous drips, respirators, and surgical wounds all breach the body's first line of defense against

ABOVE Tim Streatfield, brave survivor of the Paddington rail disaster.
OPPOSITE The innocent-looking enterococcus – in this form a bacteria that has developed resistance to almost all antibiotics.

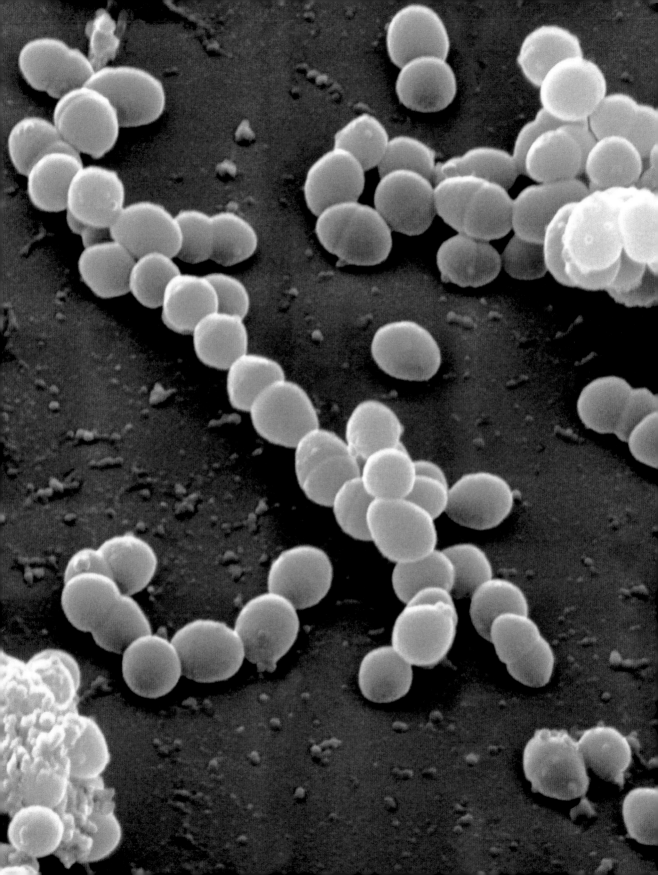

drugs—the skin and the mucous membranes. Tim's defenses were crippled. With around half the skin covering his body gone or badly damaged, there was little to protect him from the nasties in the outside world. And the "outside world" of a modern hospital provides an ideal environment for all kinds of infectious bugs to flourish. Modern hospitals are not always safe for humans.

It's uncertain just where Tim picked up the *enterococcus* that almost killed him. This particular variety of *enterococcus* is resistant to a wide range of antibiotics. In particular, it has developed resistance to a drug called Vancomycin. Until recently, Vancomycin was rightly regarded as one of the most powerful and effective of all antibiotics. But there is little doctors can do to easily kill Vancomycin-resistant *enterococcus* (VRE) once it starts vigorously multiplying.

It is possible that Tim was colonized by these bacteria at St. Mary's Hospital in Paddington, where, after the crash, he was taken along with some of the other injured commuters. But he may have picked up VRE in the Burns Unit in Essex, where he was transferred. Wherever he picked it up, infection with resistant bacteria like this is becoming increasingly common.

> … a modern hospital provides an ideal environment for all kinds of infectious bugs to flourish.

Bacterial resistance to antibiotics is one of the most important medical problems of our time. It is difficult to recall just how serious a simple infection used to be before antibiotics were available just 60 or 70 years ago. A simple cut or scratch on a finger could cause fatal septicemia. Wound infection after the simplest surgical procedure was extremely common, and often untreatable, leading sometimes to deadly results.

Even as recently as 1970, when I was a junior hospital doctor, I remember whole wards of women with minor infections being isolated after the delivery of a baby, simply because of the terrors of puerperal infection. And, until the antibiotic era, patients with tuberculosis used to be isolated, sometimes for years, in sanatoria, where either they eventually died, or the infection burned itself out and just healed, usually leaving terrible internal scarring. Syphilis would continue inexorably till its advanced stages, ineffective arsenical drugs making the patient feel even more ill, when eventually the unfortunate victim would end up insane and unable even to walk.

The development of penicillin was truly one of the great events in medical history, absolutely justifying the Nobel Prizes awarded to Flem-

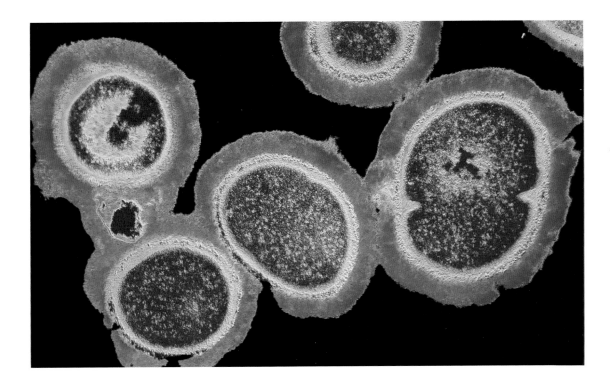

ing, Florey, and Chain. But the writing on the wall was signaled by the discovery of penicillinase, an enzyme that occurred naturally and destroys the antibiotic. Resistance means simply that the bacteria cease to be killed or inhibited by the drug. They go on multiplying and producing their deadly effects whether the drug is present or not.

We all carry this bacteria on our skin. Staphylococcus aureus *is a quiet bug much of the time but it causes boils and skin infections and can be a huge problem after surgical operations.*

Essentially, all antibiotic resistance has a genetic basis. Some bugs are inherently resistant to antibiotics, this innate resistance probably evolving because of exposure to naturally occurring antibiotics in the environment. When an antibiotic is given to a person, the susceptible bacteria die rapidly, but, equally rapidly, those with innate resistance multiply unchecked.

The situation is made even worse because bacteria can acquire resistance to antibiotics. During their rapid reproduction—a new generation of bacteria can be produced every 15 minutes—they can undergo mutation. Human evolution has taken a very long time because we only have one generation every 20 years or so. There would have been about 25,000 generations in the half million years since *Homo sapiens* first started walking in the Great Rift Valley in Africa. The same number of bacterial generations can occur within just eight months.

But being simpler organisms, bacteria mutate at a much faster rate than humans, and new resistant genes occur with frequency. Giving an antibiotic to a pool of bacteria where some of those bacteria have genes that cause resistance facilitates selection pressure. The antibiotic kills all the susceptible bacteria, selecting those that are resistant. In this way, a previously minor strain of the organism becomes dominant and potentially immune to our modern medicines.

Remarkably, there are other ways, which have only been recognized recently, that bacteria can become resistant. These wily organisms can actually transfer their genetic material to each other. Small pieces of DNA carried within the bacterial cell, so-called plasmids, can be transmitted between bacteria during a kind of bizarre "kiss." And if these plasmids carry messages for resistance, that trait is also transferred. Pieces of DNA can also be carried between bacteria on viruses that live inside them—so-called bacteriophages. Moreover, this transfer of genetic material seems to be possible between bacteria of totally different species. Frightening to think of, some single viral phages can enter a bacterium and replicate 100 copies within 20 minutes. Such a bacterial cell then breaks up, releasing many new viral particles ready for further infection.

In order for an antibiotic to work—that is, to kill a bacterium—it needs to enter the microbe and attack the target responsible for the microbe's metabolism. There are basically five ways in which the bacterial genes can resist being killed by an antibiotic. They can inactivate the drug before it gets into the bacterium. Secondly, some genes can also strengthen the wall of the bacterium so that it becomes impervious to the antibiotic. Another mechanism is that other genes may resist by pumping the antibiotic out of the bacterial cell, once it has gotten in. A fourth way resistance can occur is by the bacterium changing the antibiotic target. Sometimes this sensitive target inside the bacterium, which the antibiotic needs to destroy in order to produce its valuable effect, may become genetically altered. In these circumstances, the antibiotic may get inside the bacterium, but then "lose its way" because it cannot "recognize" its site of action. Finally, the bacterium can acquire an alternative method of metabolizing, so that the antibiotic target becomes resistant.

The ability of bacteria to evolve rapidly and develop resistance is why overuse of antibiotics is so dangerous. Killing bacteria can only make them stronger. New and more powerful strains will arise, strains that render our current antibiotics useless, forcing us to develop new ones, which in turn apply even more selective pressure on the microbes and make them stronger still. We are starting to feel the effects of these new "superbugs." In the US a staggering 90,000 patients die each year from hospital-aquired infections. Such infections strike an estimated 100,000 people in the UK per year, killing at least 5,000 of them. They account for half of the major complications of hospitalization, far out-stripping such factors as medication errors.

In Tim's case it is possible that the VRE was not picked up in a hospital. Tim may already have had VRE clinging to his skin when the two trains collided. Since VRE is normally innocuous, he would not have known he was already colonized. That would only have become evident after he had been so badly traumatized; the bug had a chance to get inside and make itself at home.

We all have bugs of some kind or another on us. For example, if we swabbed the skin and noses of 30 percent of the population, we would discover the presence of *Staphylococcus aureus*, a bacteria that can cause a range of trivial and more serious wound infections. That's fine—our skin is an effective guard against infection. But put one of those people into an intensive care unit and stick an intravenous drip in his or her arm, and we have the potential for trouble. Just a simple skin break allows these microbes to take advantage of that weakness.

The VRE clinging to Tim's skin took full advantage of his weakened state. He was kept heavily sedated for almost a month; he doesn't remember the series of near-misses he had—the bleeding in his stomach, the kidney shutdowns, the rampant septicemia. Normally, this enterococcus is not a particularly virulent organism at all. But if a patient has lost most of his skin, then it can easily find a way into the bloodstream and cause sepsis.

Tim's VRE infection was a nerve-wracking experience for Louise Teare, a consultant microbiologist at the hospital where he was treated. She admits it's the first time the staff had ever seen VRE in their hospi-tal—and it is not an easy microbe to deal with. It can fight off the effects of almost every antibiotic on the market, medicines that a few years ago would have killed off the enterococcus virus quickly and cleanly.

To treat Tim, Louise Teare and her team had to use a brand-new antibiotic—one that was so new it had not yet been put on the market. They had no experience with this drug, no idea as to what sorts of interactions to look for, and what the most likely side effects might be. But they had no choice. Without it, Tim was likely to succumb to the sepsis. And so they gave him the linezolid, the first antibiotic in its par-ticular class to be tested on humans in more than 30 years, held their collective breath, and hoped for the best.

The best seems to have happened for Tim. The drug quickly cleared the infection with no untoward side effects. But getting Tim healthy was only half the battle; the other half was trying to keep this nasty, hardy bacterial rebel from spreading throughout the vulnerable hospital population.

Trying to prevent infection on a hospital ward is an everyday procedure. Sometimes it is very difficult for health-care workers to understand why they need to carry out infection control measures, because possibly they're not seeing infections. But that simply shows that the controls are working. If we let down the guard, even a little bit, those opportunistic bugs will come crawling right back out of the woodwork.

That's why no item used on one patient on the burn unit is used on another without first being decontaminated. And that means everything. Thermometers and lunch trays are just as likely as a needle or a bandage to pick up a stray bacterium or two. Human body parts are no less vulnerable. And that's why every time a nurse has any sort of contact with a patient, he or she needs to wash in antiseptic solutions before moving on to the next patient.

And all of us are releasing skin cells into the air all of the time—cells that are likely to carry a few bacterial passengers. They may become airborne and then eventually land on any object in the room. As proof, microbiologists, such as Louise Teare, have been able to grow all sorts of organisms from just about any item you could name—radiators, fans, even doctors' white coats.

When a patient such as Tim comes on to a ward, the normal infection control procedures just aren't good enough; the staff has to practice what is known as barrier nursing.

Barrier nursing means that nothing that touches Tim leaves the room without being disinfected. When entering his room, nurses and physicians do a complete clothing change, and when they leave the room they change again. Visitors, too, have to wear gowns, and wash up when entering and exiting. And Tim himself never ventures beyond the room's perimeters. It's frustrating, but necessary. Barrier nursing protected almost all the other patients in the hospital from the infection that nearly cost Tim his life.

Despite the staff's best efforts, the VRE did manage to make its way to one other patient, causing another case of septicemia. But even with disappointment for Teare, she says it could have been much worse.

Tim is not quite out of the woods yet. The septicemia has been controlled and he is no longer septic. Teare can no longer grow VRE from swabs from the surface of his skin. However, she suspects that Tim is still carrying VRE in some part of his body— probably the gastrointestinal tract—and he's likely to do so for some time. What this means is that if Tim were to get sick or injured again—if he were to require another invasive procedure, another intravenous line—he would again be vulnerable to these potentially devastating bugs.

We have to recognize that modern pharmaceuticals have been outwitted by evolutionary cunning. Who would have predicted that in the year 2000 tuberculosis would infect one in three people on this planet? According to the most recent World Health Organisation statistics (1997), it is responsible for 2 million deaths worldwide each year, and a significant proportion of these cases are multi-drug-resistant strains that are completely unaffected by the traditional cures.

Superbugs are demonstrating that infectious disease is no longer a problem restricted to developing countries. In New York, in the 1980s and early 1990s, an outbreak of multi-drug-resistant TB took everyone by surprise. The city was completely unprepared. There was panic, especially among the HIV-positive population, who were particularly susceptible to contracting the disease, and in a greater danger of dying.

There was only one way to deal with the new strain, and that was by taking a massive daily cocktail of toxic drugs, antibiotics, and injections in the

Tuberculosis is a huge problem in most parts of the developing world. This illegal immigrant from Mexico was detained in California where he had extensive treatment in isolation for advanced TB of both lungs.

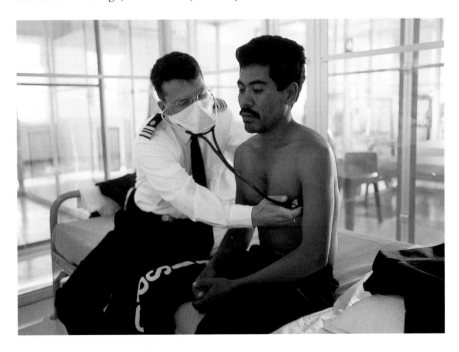

hope that the combination of toxins would overwhelm the infection. Unfortunately, this treatment also overwhelmed many of the patients. They were told to continue with the course for up to two years, but the drugs had so many unpleasant side effects that many patients refused to finish the course.

The authorities were desperate; they needed to control the disease by whatever means necessary. Their solution was somewhat medieval. They forcibly detained sufferers in a hospital called Goldwater Memorial Hospital, built on a small island off Manhattan. The patients were imprisoned there until they had finished their course, or were completely cured.

> The golden age when antibiotics and vaccination were medical miracles, and saviors of the world, is long gone.

These modern-day leper colonies could well become a more common phenomenon if superbugs become more prevalent.

The sheer vigor of these new resistant strains has left us wondering where it all went wrong. It is a humbling experience. The golden age when antibiotics and vaccination were medical miracles, and saviors of the world, is long gone.

Salome has survived being a prostitute in Nairobi for 20 years. She, and women like her, have been a key help in the understanding of HIV.

Majengo is a nasty little district in the suburb of Pumwani, near the center of Kenya's capital city, Nairobi. Much of it looks as if a puff of wind would reduce it to filthy debris. It's a place dedicated to the brewing of liquor and the selling of women's bodies, the kind of town where even the prostitutes don't venture out at night.

Salome has been a prostitute in Majengo for nearly two decades. She would desperately like to stop. She would like to sell shoes, she says, or secondhand clothes. But she has yet to find a way out. Her daughter, Mastidia, followed in her footsteps just a few years ago. She, too, would like a better life for herself and her two boys. She, too, has the pathetic, forlorn notion that one day she will better herself by going into business. But, as always, there seems no way out for these women who are still selling sex. And each time they do it, they know they are putting themselves at a fatal risk.

Like the rest of Nairobi, Majengo is besieged by a virus—the human immunodeficiency virus (HIV). The Kenyan Ministry of Health has calculated that 1.9 million of the 39 million people in Kenya are infected with HIV. We filmed in an orphanage about 50 miles from Nairobi. There are around 350,000 such orphans in sub-Saharan Africa whose parents died of the disease and who are themselves infected.

Epidemiological studies show that about 15 percent of sexually active people in Nairobi carry the virus. Frank Plummer, researcher into sexually transmitted diseases, points out that in some other parts of Kenya, the prevalence is as high as 30 percent. Moreover, the data shows that the situation is worse elsewhere in Africa, where sometimes 50 percent of the population carries the disease.

For the prostitutes in Majengo, the numbers carry a kind of irony. The going rate per client is the about 50 cents (US) for each of the six to seven clients a prostitute services on an average day. Most of the clients refuse to wear a condom. And of those six or seven men, at least one or two will almost certainly be HIV positive. Among the long-term sex workers in Majengo, the prevalence of HIV infection may be as high as 95 percent.

But what is extraordinarily interesting is that a small number of these prostitutes apparently fail to show any sign of infection, even though they are just as sexually active as their colleagues. The story of Majengo is not quite typical of African AIDS, not just a story of pain and suffering and untold numbers of sad-eyed orphans, although all of these things undoubtedly exist there. Instead, Majengo's story is about resistance to one of the world's greatest scourges. Seemingly, here at least, one catastrophic virus has met its Superhuman match—and has been resisted. But it is also a cautionary tale, the story of the remarkable resilience of the enemies

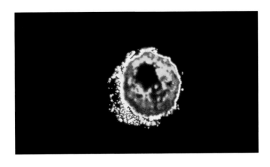

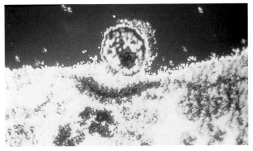

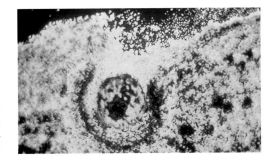

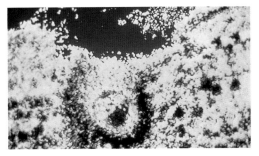

The Human Immunodeficiency Virus (HIV) photographed in the process of invading the very cells that would normally protect us against nearly all other infections. It is this ability that makes it such a deadly enemy.

with which we engage in battle, and a reminder of how we can never ever afford to let our guard down.

Many scientists in the past have assumed that a pathogen, left alone to do what it wants to do, will evolve and be less dangerous and deadly, less virulent. After all, they reasoned, what sense does it make for a bacterium or virus to kill off the body in which it lives? If we go, the microbes we carry are likely to go with us.

The problem is, that's not quite the case. Infectious diseases are not like cancer. Cancerous cells do, indeed, die when we die because they remain within us, but the pathogens that cause infectious diseases are built to disperse. An individual microbe might make its home in a single person, but its progeny are sent out into the world beyond. And so, as researchers have come to recognize in recent years, a single bacterium is willing to risk building its own tomb as long as its offspring, carrying its genes, its infectious heritage, manages to find another host in which to begin the cycle yet again.

I spent an afternoon face-painting many of these delightful children. All are orphans, all have HIV, and many have tuberculosis as well. The child sitting on my lap and the two boys next to her are now dead.

And that is just what happens each time a person with a cold starts to cough or sneeze, or a cholera victim's feces are washed into the water supply, or an AIDS patient has unprotected sex.

That is not to say that pathogens don't care about our survival. The longer we are alive and kicking, the longer they can create new generations of infectious offspring. But, as is the case with any other evolving living thing, there are evolutionary trade-offs that determine how dangerous a bacterium or virus can be.

Case in point: The common cold virus, also known as the rhinovirus, does not often kill its human host because there is no need. For one thing, it's usually the only infectious agent in a given human body. It has no major competition for the space and resources it needs to make many copies of itself and send them out into the world. Instead, it simply needs to push the body to produce lots of virus-carrying mucus. If a cold were to make you so sick that you couldn't get up out of bed, the cold virus would have a much harder time making its way to new victims. Instead, a rhinovirus will get the most mileage out of you if it keeps its wrath in check, and allows you to interact with—and infect—as many people as possible.

HIV can lie low for years, without having a single deleterious effect on the health of the carrier. The virus is in no rush to replicate and infect the cells in our bloodstream. During this time, on average eight to ten years, the carrier may be completely unaware that he or she is carrying the disease, thus providing an ample window of opportunity for the virus to be transmitted to others. It is this long gestation period that has caused the most damage, and the prostitutes of Kenya who carried HIV for years were the unwitting allies of this terrible disease.

Salome and Mastidia come to the clinic in Pumwani once a month or so to have blood tests and to pick up a supply of condoms. Dr. Kimani, a senior physician, estimates that the clinic gives out over ten million condoms each year. And when prostitutes come for checkups, the staff does its best to make sure the women leave understanding a little better how to protect themselves. "We've made them feel part and parcel of the clinic," says Dr. Kimani. "We've made them feel that the clinic is theirs."

The workers at the clinic have established community outreach programs, which are largely peer led. They talk to the women about the facts of life in a world riddled with AIDS. "If a person's blood is infected, it stays that way," a nurse tells them at an informal meeting where the women can discuss their worries. "It cannot change back to being clean. If it is clean, it can become infected, but if you have been told it is infected then there is nothing you can do."

They talk to the women about the importance of good nutrition, about taking care of themselves. They talk about the importance of seeing a doctor as soon as they feel ill. And, of course, they talk about protection—about using condoms with each and every customer—and press condoms into the women's hands. They tell them to insist. They sometimes even get the women to share the names of clients who refuse to use condoms, creating a blacklist of sorts. They talk to the women about how serious all of this is, telling them it's a matter of life and death.

Salome and Mastidia have not always listened to this sort of talk. Salome was a prostitute well before AIDS was taken seriously in this part of the world; the men with whom she did business would never have dreamt of using a condom. And while both women claim they now use one with every customer these days, Dr. Kimani knows that they are unlikely to be telling the truth:

"I would say 90 percent have a regular client. And this regular client is sort of a boyfriend. And what happens is, he kind of pays, yes, but he pays in kind. He offers protection, because it's a slum area and it's risky and there's a lot of mugging and a lot of fights. And they need somebody to protect them. The only problem is that they will insist on condom use with the other six to seven clients, but their regular client is kind of given some privileges. And those privileges are that he won't use a condom."

Mastidia has just such a "regular client"—and he is HIV positive. But, extraordinarily, Mastidia appears to have remained free of the virus. So has Salome. Despite years of prostitution and untold numbers of unsafe encounters, Salome and Mastidia have repeatedly tested negative for HIV.

The apparently privileged position of these two women is not that unusual. Dr. Plummer has been conducting a follow-up on women like Salome and Mastidia since 1984, when he first came to Kenya to study the spread of gonorrhea infection. Salome was one of the women in his original study of a group of prostitutes who were willing to have their blood tested in exchange for free medical care in a modern facility. He started his researches when HIV was just being recognized as a problem in Africa. But no one then seriously believed that HIV would be much of a problem; indeed, there was some doubt that the virus would be found

> There is a subgroup of women … who, despite many years of heavy exposure to HIV, don't get infected … In some way … their bodies have managed to resist what has always been thought of as an irresistible virus.

at all. When they tested 500 of the women, they were shocked to find that 65 percent of them were HIV infected. It was obvious that HIV was an enormous problem.

Despite the initial shock over the extent of HIV infection in those relatively early days of the epidemic, the real surprise has continued to be the number of women who remain virus-free. "There is a subgroup of women in the cohort who, despite many years of heavy exposure to HIV, don't get infected," says Plummer. "We can't find HIV in their blood, but know from epidemiologic studies that, in a statistical sense, they should be infected. They should have had enough exposure."

Dr. Kimani points out that, even with the most conservative estimates, these women are likely to have been exposed to HIV 70 to 80 times each year. There really is no statistical likelihood of them merely being lucky enough to avoid infection after such persistent and repeated exposure. Undoubtedly, there is something very remarkable about these women. They have encountered this deadly virus, but apparently fought it off; in some way, which is not understood, their bodies have managed to resist what has always been thought of as an irresistible virus. There would appear to be a great deal we could learn from studying them.

Today, 2,000 prostitutes are being studied by Dr. Kimani. Many of them, especially the ones who joined the study in its early days, are dead or dying. Indeed, says Kimani, in less than a decade of involvement, he has seen more than 700 of these women die of AIDS-related illness.

Despite these deaths, there is hope among the women who are infected and for their health-care workers, hope that springs from no fewer than 140 of the women who have shown a continued resistance to HIV. For the people at the Pumwani clinic, this extraordinary fact has almost become their *raison d'être*: to find why these women can beat the human immunodeficiency virus. With that understanding, we might be able to use the knowledge for the advantage of everybody. With more information, a vaccine may be a possibility—vaccination would be a method of providing immunity to the disease, rather than focusing on attacking the virus once it has taken hold, a goal which is still out of reach.

To those who are skeptical about an HIV vaccine, Dr. Plummer points out that the eradication of smallpox began in a somewhat similar fashion. In the 18th century, Edward Jenner discovered that some milkmaids appeared to be immune to the killer infection, smallpox. Those who had been infected earlier by the relatively benign virus that causes cowpox were, it seemed, protected against smallpox. "In some ways," says Dr. Plummer, speaking of the prostitutes, "these women are the modern equivalent of Jenner's milkmaids."

The principle of immunization has not changed since Jenner first discovered it. This involves the intentional infection of a patient with a less virulent form of the virus or bacterium, in order to ready the immune system in preparation for the full force of the disease.

A certain kind of white cell, called a memory lymphocyte, remembers the makeup of the microorganism contained in the vaccine—and this extraordinarily robust cellular memory can last a lifetime. These white cells are able to rustle up a very speedy response if a more virulent infection of the same type were to take hold, and antibodies can be quickly produced to stop the infection in its tracks.

Studying these modern-day milkmaids closely has turned up some interesting twists to the story. Normally the HIV virus infects T cells, cells that have a commanding role in the body's immune system. One might think that Salome and Mastidia might have T cells that are somehow protected from the virus, but it turns out that the women do not have virus-proof T cells. If one takes the white cells from a sample of their blood and exposes them to HIV, those cells become infected. In the body, however, something keeps the virus from setting up shop. Somehow prostitutes such as Salome and Mastidia have acquired immunity to HIV. Dr. Plummer explains: "What's happened, I think, is that they've been exposed to the virus, had a low-level infection, and then—because of characteristics of their immune system that we don't understand—were able to get rid of it, and are now protected, at least to some degree."

In other words, these women are armed to the hilt with defenses that the rest of us are generally unable to muster. Circulating in their bloodstreams are cytotoxic T cells, a different kind of white blood cell that is able to recognize and kill those T cells that are infected with HIV, thus stopping the virus from replicating. In addition, they produce large numbers of helper T cells, which prompt the production of additional cytotoxic cells when necessary. Finally, they have antibodies to HIV—proteins that attack the virus—in their genital tracts, where the virus is most likely to enter.

Plummer has not found antibodies in their blood, but he has found them in the vagina and cervix. These antibodies can neutralize HIV before it even enters the bloodstream. So it seems a combination of factors has given these women an immune system that is prepared for the worst, and able to deal with it.

Having uncovered the source of the immunity, Plummer and Kimani are now looking for its possible genetic basis. According to some statistical data collected by Kimani, a fairly large percentage of the approximately seven percent of sex workers who are immune to HIV are related in some way. It seems that women who are related to a woman who is immune are about twice as likely to be immune to HIV as those who are not. So, perhaps, this genetic component may provide some answers as to why some immune systems can cope and others cannot.

To investigate this enigma, the Kenyan researchers have hooked up with a team from Oxford under Professor Andrew McMichael (an old school friend of the author) and a Canadian team based at the University of Nairobi. They hope to work out how to mimic the immunity found in the prostitutes. At present, they have evidence of an extraordinary phenomenon, but no one knows whether this knowledge will eventually translate into an effective vaccine for HIV.

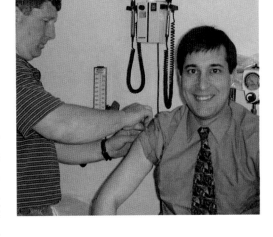

During filming in Oxford, I was able to see the Oxford team at work trying out a new vaccine based on this curious immunity. One volunteer who elected to have an injection to see what the body's response might be was Dr. Evan Harris, a member of Parliament. Great credit is due to a politician who gets himself involved in a trial of this kind and supports the need for better medical care in Africa.

Dr. Evan Harris, a member of Parliament, doing his bit for British science and for the children of Africa.

But there appears to be a deadly irony about the apparent immunity of these women. Ten prostitutes who had remained HIV negative in spite of the most persistent exposure became HIV positive after they had taken a break of a few months. It seems that those prostitutes who remain healthy while being exposed may need their immune system to be constantly and repeatedly primed. Giving up "the life" may prove to be the death of these women.

Even the most outlandish imagination could not have predicted the terrifying rise of HIV, its stealthy spread across the world, and our powerlessness in the face of it. But because of these fears of HIV, virulent flesh-eating bugs, and resistant strains, we tend to forget that the most important relationships we have in our lives are with the bugs we harbor inside our bodies. There are far more bacteria in our bodies than there are cells. We are quite literally covered in them. Each square inch of our skin is a home to some 10,000 bacteria. These microbes have a key role in preventing infection, and bacteria in our gut help us digest our food. Even those that are potentially harmful are dealt with, most of the time, by an immune system that is well used to microbial invasion.

The fact that we have lived and died with these tiny organisms for millions of years is something we tend to lose sight of. We chase after microbes with a vengeance, without wondering whether we may have gone too far. These days we not only try to protect ourselves from chance encounters with potentially deadly microbes, but we also try to eliminate the common cold and send the flu virus into extinction. We clobber even the simplest infection with heavy-duty antibiotics. We preempt a whole range of infectious diseases with large scale vaccination.

We have done all this without understanding the long-term effect on human health and well-being, but we are now beginning to find out. Is our eagerness to kill off these bugs misplaced?

Antibiotics do not just kill dangerous bugs, they often kill the helpful bacteria, too; this means that our defenses are lowered, and secondary infections can set in more easily. Secondly, heavy use of antibiotics is very likely to provoke resistant strains to appear. These are the dreaded super-bugs for which we need to develop even more powerful antibiotics.

> Allergies are truly modern afflictions. Cavemen didn't have allergies…

The third reason is a side effect that no one seemed to predict. Many people think that the plagues of the 21st century will not be variants of Ebola, HIV, or some other new and fatal super-bug. They think the new plagues could be allergies, such as asthma and allergic rhinitis (hay fever), which have increased in the past 20 or 30 years to startling proportions. We are just beginning to find out why.

Allergies are truly modern afflictions. Cavemen didn't have allergies, or, at least, probably only very rarely. Today, between 30 and 40 percent of us will test positive on a skin test for reactions to at least one allergy. And five to ten percent of adults in the Western world will experience those reactions escalating into the chest-tightening, breath-stealing condition we call asthma.

The best definition of allergies is "an inappropriate response of the immune system to environmental allergens." A normal, healthy immune system is not programmed to respond to house-dust mites, pollen grains, or cat dander. These objects are not dangerous—in fact, they are completely benign and should elicit no immune response at all. But for people who are predisposed to developing allergies, the immune system responds out of proportion to the threat, producing mediators that then result in inflammatory reactions including asthma, hay fever, and eczema. In other words, we are reacting to a nonexistent threat; our body is turning on itself.

Today, about 17 million Americans have asthma. Nearly one-third of them are under the age of 18, and collectively, they are absent from school more than ten million days a year. In the UK, nearly three million people have been diagnosed with asthma, including one in every seven schoolchildren. The rate of asthma in the developed world is doubling every 20 years.

We already know that both allergies and asthma have a genetic component. If one of your parents has an allergy, you have a 25- to 30-percent chance of turning up allergic yourself. If both your parents are allergic, your chances rise to 60 percent or more. However, genes could not possibly explain the rapid rise in asthma over just a few decades; they simply could not spread throughout the population that quickly.

So, if the problem is not in our genes, it is likely to be in our environment. But what could be causing us to respond to normal, nonthreatening

We try to protect our children by stopping them from eating dirt, but this strategy may actually cause considerable harm. Modern allergic diseases such as asthma may be caused partly by such overprotectiveness.

particles, such as pollen, at an ever-increasing pace? What could precipitate such a fundamental breakdown in our immune system?

There is a popular notion that modern pollutants, such as gasoline fumes, additives in processed foods, and other synthetic chemicals in the atmosphere that we inhale or ingest on a regular basis, can cause allergies. The "boy in the bubble" case was a terrifying harbinger of what a pollutant-laden environment can do to human health; it appeared that he was, to all intents and purposes, allergic to modern life.

This idea has been discredited. There are certain chemicals, such as sulfur dioxide, which can provoke already existing allergies. But for the most part, pollutants do not precipitate dangerous acute allergic reactions. The boy in the bubble turned out to have an inherited genetic condition, not a series of allergies. Other cases have been shown to be psychosomatic—the patient believes so intently that he or she is reacting that it becomes a self-fulfilling prophecy. Other cases have been shown to be frauds.

Perhaps the problem is something that *isn't* in our environment. An increasing number of scientists are saying that we have deodorized, sanitized, and homogenized our way to this allergy epidemic. We are, to put it simply, too clean. This theory is called the Hygiene Hypothesis.

> An increasing number of scientists are saying that we have deodorized, sanitized, and homogenized our way to this allergy epidemic.

One proponent of this theory is Ratko Djukanovic of the University of Southampton in England:

"I like to think of this hygiene hypothesis as the immune system becoming mischievous, much like a child becomes mischievous when it doesn't have anything to do. Instead of reacting to things like bacteria and viruses that are normally present in our environment, because we live in a healthy—not to say sterile—environment, the immune system looks for other things to do. And what it does, then, is it starts to respond to things like allergens, pollens, house-dust mites."

A stream of research data from around the world—Africa, Germany, Sweden, Italy, Israel, and the UK—begins to back up what, at first, was nothing more than an intriguing idea. Scientists are finding that children exposed to more infections when they are young—children with older siblings, those who go to day care at an early age, those who live in less than hygienic conditions—have

OPPOSITE A dust mite, looking more menacing than it normally appears. Invisible to the naked eye, this unfriendly little creature is responsible for many modern human allergies.

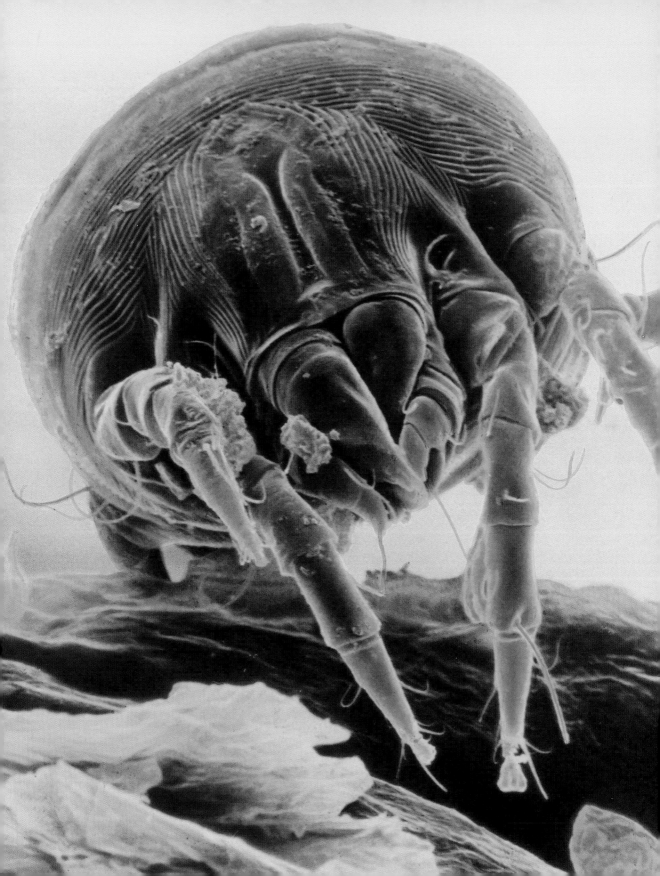

much lower rates of allergies, asthma, and autoimmune diseases, which seem to be the result of an immune system with nothing better to do.

At its core, the problem might be a legacy of our evolutionary past, an imbalance involving an arm of the immune system that originally developed to help us fight off parasitic worms. Now that worms are of little concern in Western society, these so-called Th2 T cells are instead responding to allergens such as cat dander and ragweed pollen. They are sending out a chemical signal that sets off the inflammatory response that is the hallmark of allergic reactions.

In most of us, the number of Th2 cells, or parasite-killing cells, is low; instead, we have a preponderance of Th1 T cells, which are the cells responsible for responding to pathogens such as bacteria and viruses. But in people with allergies and asthma, the Th2 cells are in charge, and the results can be debilitating. With no parasites to deal with, they look around for something else to attack.

Djukanovic says that in our immune system "there should be a balance—if you will, a yin and a yang—between the mediators that protect us against infections and those that were meant to protect us against parasites, but which now respond to allergens." For those who suffer from allergies that balance has swung toward the Th2 cells.

So why are more and more of us living in Th2-dominated bodies? The supporters of the hygiene hypothesis think the answer is all around us. There have been many changes in our environment. Houses have become warmer, and this has allowed for more allergens in our households. We keep more pets; even if you don't have pets at home yourself, children cannot go to school without being exposed to them. We have made our environment cleaner.

The crux of the argument, then, is that we now have reduced our exposure to infections, through an almost sterile existence on the one hand and use of antibiotics and vaccination on the other. Developing countries do not have an allergy epidemic. The reason is probably that conditions are far less antiseptic. Kids play barefoot in the dirt. They spend more time together at an earlier age, thus spreading infection more quickly.

And, according to the hypothesis, stimulating the immune system with bacteria is good for you. Of course, no one wants to encourage serious or fatal infections—there is certainly more than enough of those in the third world at present. But Djukanovic says that, perhaps, we should encourage a child to play barefoot in the dirt and maybe even taste it. We should not go overboard with antibacterial toothpaste, hand-washing, or dishwashing liquid.

"What we're suggesting is that, perhaps, we've overdone it," Djukanovic says. "We've taken away all the bacteria that may be necessary, in fact, for us to maintain the balance of the immune system." In 1997, a team of Japanese scientists published a paper in the journal *Science* showing that asthma rates were significantly lower in schoolchildren who had been exposed to and had fought off the bacteria that causes tuberculosis. At around the same time, Djukanovic and his colleagues began to hear talk about scientists from University College in London who were developing a vaccine using *Mycobacterium vaccae*, a nonpathogenic cousin of *Mycobacterium tuberculosis*, the very microbe that causes the disease of the same name.

Mycobacterium vaccae, found in the African continent soil, was carefully studied in the laboratory before it was ever put into a hypodermic needle. What the researchers saw gave them hope. It seems that *Mycobacterium vaccae* stimulates just the T1 arm of the immune system and suppresses the allergy-provoking T2 cells and their mediators.

Djukanovic's team and a team from University College decided to test whether this vaccine might be able to change the response of asthmatic patients to allergens that normally set them wheezing. In the first phase, they recruited 24 young men, most of whom were sensitive to the common house-dust mite—the most common allergen in the UK (and in the United States as well)—and all of whom had asthma. Twelve of the men received the active vaccine, while another twelve received a placebo.

The patients who participated in the trial had an allergen test, and then inhaled an aerosol of the allergen to which they were sensitive. This caused a mild asthmatic reaction, which was recorded in the laboratory. And then they were given the vaccine. The patients returned after three weeks and the process was repeated, after which experimenters compared the magnitude of the response to the inhaled allergen before and after treatment.

Mark Dyer can barely remember a life without asthma. He was just three years old when it was diagnosed, and says the condition hit its peak when he was 12 or 13. He says the only way he can describe an asthma attack is that it feels like he is being strangled while somebody is sitting on his chest at the same time, and he is trying to breathe in but nothing happens. It can be terrifying.

Mark tries never to let it get that bad. He has been controlling his asthma since his teen years with tablets and inhalers, but a life tied to pharmaceuticals is, he says, very restrictive. And so, when offered the chance to take part in the experimental treatment, he jumped at the opportunity to be rid of the Ventolin.

For Mark, who later found out he had been given the active vaccine rather than the placebo, the treatment was almost anticlimactic: no fancy equipment, no long hospital stay, just a couple of tests, a shot in the arm, and he was sent on his way. The results of that injection, however, were anything but disappointing.

"Since the injection, there's been a marked difference," Mark says, just a couple of weeks later. "I haven't had to use my Ventolin inhaler as much, if at all. And my asthma seems to be very much more controlled. It doesn't affect me as much. And as time goes by, it's less of a burden to me. I don't tend to think about it as much."

Even better, the members of Mark's soccer team don't think about it as much, either.

"I've always loved playing soccer," he says. "I've played it from a very early age. Before I took part in the trial, the asthma played a big part. I used to have to run to the line after only 10 or 15 minutes to have a puff of my inhaler. Since the trial, I won't say my soccer has improved, but I'm less reliant on my medication and I can play the game better inasmuch as I'm not slowed down by heavy breathing or whatever. I can control myself a lot better. And other members of the team don't even notice I'm asthmatic, which is a big compliment, really."

Ratko Djukanovic and his colleagues weren't expecting much from a single dose of what they expect will be an ongoing treatment. He says they were "quite intrigued, and I dare say delighted" to find out that Mark had had such a positive response to the vaccine. Still, as he says, one swallow does not make a summer, so he needed to be absolutely sure that Mark is representative of the population at large.

Mark was not the only success story. Looking at the analyzed data from the trial, the researchers found a 30-percent reduction in the severity of the allergic response among the treated patients. Equally promising were the laboratory findings, which showed that the vaccine did, indeed, seem to be having a measurable dampening effect on Th2-signaling chemicals.

There are many more steps that must be taken before the vaccine might be approved as a treatment for asthma. The team is required to conduct a large clinical trial, involving several hundred patients, before it can open the champagne. But the results, as Djukanovic says, are "encouraging."

Moira Brown, a researcher at the Southern General Hospital in Glasgow, Scotland, did not start working with the herpes simplex virus in the hope of curing cancer. She started working with it mainly because it is stuffed full of genes. It is a useful model to understand how viruses function, what they are made out of and how the genetic code within the virus controls the infection. There's a relatively large genetic content in it compared to a lot of other viruses, which makes it easier to work on and experiment with.

The herpes simplex virus preys on nerve cells of all sorts, slipping silently into the cell's body and depositing its viral DNA into the nucleus. It then stages a cellular coup, taking over the machinery that allows the cell's DNA to replicate itself and instead pumps out copy after copy of its own strands of genetic material. Filled to bursting with new virus particles, the neuron blows apart, sending all those particles streaming forth, ready to infect another cell.

… injecting a virus into somebody's brain is a fairly radical thing to do.

There is nothing particularly unusual about this pathogenic routing. This is how almost all viruses make their living: infecting the cell of their choice, replicating wildly, and then bursting forth to infect again. What makes the herpes virus different—and very interesting—is that it does not pursue its goal quite so single-mindedly as many of the others. In fact, the herpes simplex virus often sits quietly in a neuron for weeks, months, even years. Only when the virus is shaken up—when the body is under stress, or under attack from another virus, or exposed to a lot of sunlight—does it come back to life.

This unusual property allows scientists such as Moira Brown more room in which to maneuver. It gives them the chance to gain control over the virus, to manipulate it for their own uses. And to get that control, they had to find the genes necessary for viral replication and yank them from the nucleus.

The virus can still infect nerve cells, but without these particular genes they could not go through the normal replicative process and destroy the nerve cells. Brown explains: "So that made us think, well, could this be used as a therapeutic tool? Because we have turned something that was potentially bad into something that was potentially good."

The hobbled virus had another intriguing characteristic. Normal nerve cells do not divide; in fact, some of the neurons strung through our spinal cords last an entire lifetime, and so do not need to be replaced.

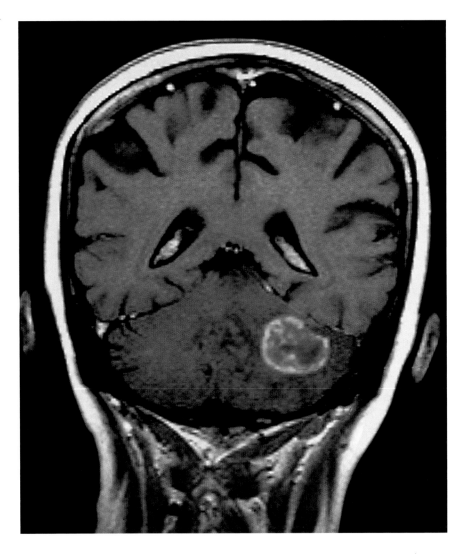

A magnetic resonance scan of the head showing a tumor in the area of the brain that controls balance.

The modified herpes virus was unable to replicate in these normal nerve cells, but it seemed able to grow quite well in cells that were very rapidly dividing. In an adult, the only rapidly dividing cells in the brain and nervous system are those that should not be there in the first place—cancer cells.

Was it possible that these mutated viruses could infect tumor cells and destroy them while leaving normal brain tissue alone? It was an intriguing possibility.

And so Brown and her colleagues began a series of experiments to explore the potential of modified herpes simplex. They grew cancer colonies in culture and then infected the cells with the virus. Just as they had hoped, the virus wormed its way into the tumor cells and then burst right out of them, infecting nearby cells in the colony while leaving healthy neurons alone. It was no less successful in living creatures, destroying a range of tumors—brain, skin, lung, ovarian—while leaving the normal tissue perfectly intact.

It was a startling and exciting realization. A new and effective therapy for cancer seemed within reach, a treatment which, unlike radiation therapy and chemotherapy, did not simultaneously destroy healthy cells. But pursuing this line of research was going to be risky. It would involve putting into the human brain a virus that can, in some cases, cause encephalitis, a deadly brain inflammation. On the other hand, if it worked, it might offer a treatment for the sorts of brain tumors for which there are currently no therapies available at all.

The UK Gene Therapy Advisory Committee took some persuading but, by the autumn of 1997, it gave Brown and her colleagues the go-ahead for a trial. The modified, disabled herpes virus would be injected into the brain of a patient with malignant glioma. (Malignant glioma is a primary brain tumor of the supportive tissues that surround neurons; there is currently no cure for it, nor even a definitive life-extending treatment.) With the surgeon guided by a three-dimensional image of the brain in order to put the dose in just the right spot, the virus was to be delivered directly into the tumor itself. It is a relatively simple procedure, but then injecting a virus into somebody's brain is a fairly radical thing to do.

The first trial was a safety trial; a small amount of the disabled virus was deposited into the tumors in the brains of nine glioma patients. It was a nail-biting exercise for both the scientists and the patients. This was the first time in the world that a virus of this kind had been tested in humans.

The trial seemed a success. None of the patients showed any adverse reaction to the virus, nor did any of them get encephalitis. There were, in fact, no toxic side effects noted at all, and, in some of the patients, the researchers even began to observe some diminishing of symptoms.

The very first of these patients was a young Glaswegian named Robert Swan.

Robert's battle with glioma began with a twitchy hand. And then he began sleeping, constantly. It took more than six months, and a number of specialists, before Robert was diagnosed with glioma. His doctor told him he had four to six months to live. His condition was deteriorating quickly. He lost movement in his left arm and leg as the tumor pressed down on

Robert Swan at a Celtic soccer game in Glasgow, Scotland.

the right side of the brain controlling their function. He was losing his sight in one eye. And then they found out about the herpes virus trial.

His mother, Marie, didn't hold out high hopes for the treatment. She knew it was highly experimental, and that Robert was to be the very first person to try it. Still, it was a chance. "And Robert wanted to do it," says Marie, smiling. "You can't say no to him—he's quite determined. At that time, they'd been in and out of his brain that much that I don't think he cared."

Robert got his dose of virus in October 1997. He was monitored carefully, kept in the hospital for about a week, and then released.

Marie kept looking for some major change, but didn't see any. Then after a few months had passed, her son was still alive—still alive months after the doctors thought he would be dead. And his tumor was no longer growing and no longer consuming more and more of his brain matter.

Today, Robert Swan's tumor looks just as it did two years ago. Since he was injected with the virus, he has had no other treatment for his cancer. His mother has also begun to see more positive changes. He can speak, she says with a palpable relief. He can remember things. His leg is getting stronger.

However, Moira Brown cannot yet determine what the state of the tissue is *within* the tumor. It's possible that, although not growing, the cancer cells are still active. Experimenters hope that the cells are completely inactive. Brown says the tumor "certainly hasn't progressed, and I think it could be said that the disease is stable at the moment."

She cannot say for sure that the current state of Robert's tumor is a direct result of the virus, at least not until she's done a full clinical trial—one that has large numbers of patients, a control group, and an impartial analysis of the data.

Nevertheless, she and the team are optimistic. After all, the patients in her first trial, as is the case with many trials or more or less untested therapies, were at the end of their tether. They had already been treated for their gliomas with every available, applicable therapy out there, and their cancers had sprung back to life once again. This initial trial, in any case, was to determine the safety of the therapy.

In this next phase of the trial, the virus will be injected into newly diagnosed glioma patients, in whom the cancer has not become quite so comfortable and is not, perhaps, quite as aggressive.

And glioma isn't the only cancer that is affected by the herpes viruses. Brown is testing the virus on patients with malignant melanoma, the most aggressive form of skin cancer.

It is the glioma trial that has her truly excited: "The amazing thing is that the patients actually did better than we could have realistically expected," she says. "The fact that four of them were still alive between one year and two years post-virus injection is, I think, amazing."

The task for modern medicine is not only to keep up with and combat the threat of infectious diseases, but also to try and find ways to use them for our benefit. Using microbes to tackle both allergies and cancer is a promising line of research. We are learning to turn killers into cures.

But in order to find new therapies for allergies such as asthma, we have had to recognize that we can go too far in eliminating infection from our lives and that we should not upset the natural balance of our immune system. Similarly, we are beginning to realize the unforeseen and undesirable consequences of overusing modern medicine, particularly the role of antibiotics in promoting resistant strains.

> Using microbes to tackle both allergies and cancer is a promising line of research. We are learning to turn killers into cures.

Infectious disease will remain a formidable enemy. A modern-day Black Death, which devastated civilized Europe hundreds of years ago, is still a real possibility—and a possibility which, if anything, is certainly more likely with time. Despite our best technological efforts, we are still vulnerable and may become more so as increasingly virulent and hardy strains develop.

This, above all, is why we must use antibiotics with restraint and not for trivial infections. Certainly, we should not expect an antibiotic every time we go to a doctor with a minor viral or bacterial infection, but allow the Superhuman within us to do its work instead. It is equally foolhardy to give antibiotics to our livestock, simply because these seem to improve their growth—and make the meat more marketable and profitable.

We live in a crowded world, a world in which millions of people travel across all five of its seven continents every day of the year, carrying infection as they do so. If a plague were to descend now on Athens, we would all reap the consequences.

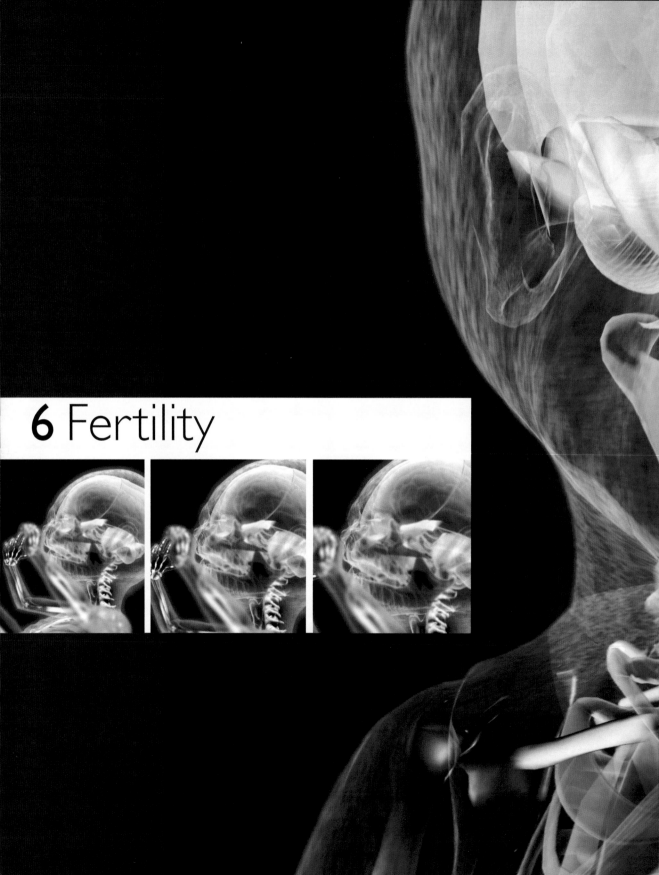

6 Fertility

The biblical story of Jacob and his untrustworthy father-in-law, Laban, is interesting for two reasons. The first reason is the unrelenting maleficence displayed by Laban, a man who might, in other circumstances, have gladly sold his own grandmother. Certainly, Laban displayed considerable greed. Rabbinical legend has it that when he kissed Jacob on meeting him, he did so because he hoped to locate jewels or money that Jacob had concealed in his clothing or mouth. Secondly, the main story strongly suggests that here was an extremely early indication that people in ancient times understood the basic principle of genetics and heredity.

Jacob worked for 16 long years managing his father-in-law's farming activities. Only at the end of his employment did Laban ask Jacob how he would like to be paid. Cannily, Jacob refused any money, requesting instead the few sheep from Laban's flock that were speckled with black spots. Laban readily agreed, at least verbally. Unlike the pure white sheep, the speckled ones were not much sought after, and, in any case, they only made up a minority of his flocks. Laban, however, was not the kind of man to hand over even a small part of his flock, even to a devoted employee and son-in-law. So, once he had agreed to give the speckled sheep to Jacob in payment, he ordered his sons and servants to remove all the speckled sheep from his flock immediately. They did, and Jacob never saw them again.

> Since biblical times, and probably before, we have manipulated reproduction …

Jacob was undaunted. In the remaining time he spent in Laban's employment, he set about breeding speckled sheep from the pure white flock that remained. But how? The Bible reports that Jacob shook speckled sticks at the sheep while they were drinking and probably while they were copulating. And there was a widespread belief up until comparatively recently that prenatal events, and even what happened at the moment of conception, could influence the physical characteristics of a child.

Of course, in reality, Jacob and his speckled sticks alone had no chance at all of increasing the chances of breeding speckled sheep. But Jacob was skilled at husbandry. He had been breeding sheep for the past 15 years and must have realized that sometimes two pure white sheep can give birth to a speckled lamb.

Moreover, he would also have observed which of the pure white sheep were capable of producing a speckled lamb. The speckling is almost certainly what we now know to be a recessive genetic characteristic; an animal can be a carrier of the speckling gene, but not be speckled itself.

Jacob Setting the Peeled Rods before the Flocks of Laban by Bartholomé Murillo, 1617-82. To this day there still is a notion that events at the moment of conception may influence a child's characteristics, but Jacob's purpose here was surely hiding his understanding of genetics.

Therefore, if two carriers of the speckling gene produce a lamb, there is a chance that their offspring will have both copies of the gene, and therefore the sheep will be speckled.

Although he may not have known the genetic mechanisms of recessive characteristics, Jacob knew enough to produce a massive flock of entirely healthy speckled sheep. It is relatively simple to calculate. Knowing which white sheep carried the speckling trait, it would be possible to ensure that every single animal born within three generations of starting this exercise would be speckled. Laban, furious, had no choice but to hand the sheep over. Jacob's flock flourished and he eventually became far richer than his father-in-law.

Since biblical times, and probably before, we have manipulated reproduction, and not just in animals. There are, for example, references in Vedic literature, the most ancient sacred writings of Hinduism, to the possibility of increasing one's chances of conceiving a boy or a girl. Fertility treatments also go back to ancient times. Of the four matriarchs, Sarah, Rebecca, and Rachel were infertile. Only Leah, Jacob's unloved first spouse and ironically the daughter that Laban had tricked Jacob into marrying, was highly fertile. The three infertile matriarchs were said to have tried various remedies for infertility. At least one of them used mandrake, a poisonous plant with narcotic properties. The huge forked root of this plant was thought to resemble the lower half of the human form, which may have been why it was associated with fertility.

Modern fertility medicine continues to generate its fair share of myth. Fertility is a source of great emotional stress and over the years, a number of practitioners have cashed in on people's fears and desires surrounding conception and pregnancy. Many treatments that have no more basis in science than the root of the mandrake have been peddled.

But now we do have a very real ability to intervene in the process of conception and the genetic legacy that we pass on to our children. In vitro fertilization (IVF), together with molecular biology, has transformed the whole field. Our knowledge of genetics and DNA has exploded since Frances Crick and James Watson first stared at their model of the double helix with the absolute conviction that they were right. We have a much improved understanding of the processes of cell growth, division, and fertilization and this knowledge has given us the means to manipulate fertility like never before. Medicine is interfering earlier and earlier in life; we are now able to meddle not only with the developing embryo in its mother's womb but with the process that starts the beginnings of life itself.

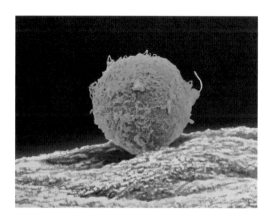

ABOVE *A colored Scanning Electron Micrograph (SEM) of sperm clustered around a human egg or ovum during fertilization. The egg (blue) has fine, hair-like sperm (green) attached to its thick, spongy surface. Only one of the millions of male sperm may penetrate the egg's wall to fuse with the nucleus.*
OPPOSITE *Another electron micrograph showing a human egg magnified 3,000 times. The remains of cells (colored yellow) from the ovary, from which it has recently been ovulated, are stuck to the surface.*

In theory, at least, each of us is born with a prodigious ability to procreate. From puberty onward a man produces half a billion sperm each day. Each time he ejaculates he is likely to expel between 100 and 300 million sperm cells. It just takes one single sperm to fertilize an egg.

A woman is a little less of a reproductive overachiever. Midway through her development, a female fetus will have produced seven million eggs in her ovaries. By the time she is born, that number will have dropped by more than half to below three million. And this death march continues: When she reaches puberty, a teenage girl is carrying less than 500,000 of these cells.

The numbers continue to drop throughout the rest of her life and not just because of those she ovulates. Attrition actually takes the majority of eggs; they simply seem to die within their follicles. And by the time menopause arrives, her ovaries will be largely empty. Many other animals, such as

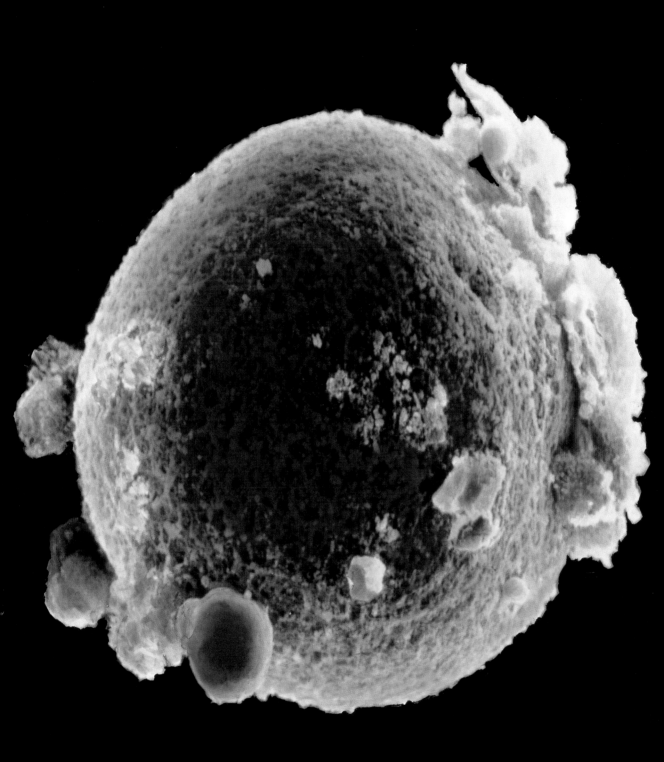

mice, also show gradual depletion of their ovaries as they age. However, they do not seem to experience menopause as such. Menopause is rare among other mammals, probably because most animals do not have the capability to protect themselves from the environment that surrounds them and they die before the effect of ovarian aging can be seen.

For the average woman, half a million eggs would seem to be plenty, considering that she needs to release only one egg every month from puberty until menopause, except, of course, for any time during which she is pregnant. On average, then, a woman liberates no more than 400 to 500 fully mature egg cells, in total, in her lifetime. But these numbers are flawed; the real situation is actually a good deal more complicated than it seems.

Although only one mature egg is ovulated each month, many more eggs are initially started on the maturing process with each menstrual cycle. Perhaps as many as 10, 20, 30, or even 40 follicles containing eggs are stimulated to commence growth, but in humans normally only one follicle becomes fully ripened and ruptures to release its egg. More than that and twins would be much more common than they are. The other follicles undergo a process called *atresia*—they wither, and the egg inside each of them undergoes programmed cell death. Just why this happens is not clear; there certainly seems to be no obvious evolutionary advantage.

Nevertheless, each ovary contains a huge surplus of eggs that do not appear to fulfill any requirement. With such overwhelming numbers of potential germ cells, one would think it would be difficult to avoid becoming pregnant, yet, even in couples who are normally fertile, intercourse in any given month offers little better than a one-in-five chance of conceiving a child.

This natural low fertility is probably just as well. If humans were more fertile, the planet would have become far more overcrowded even earlier and it is difficult to imagine how the human race could have survived the massive depletion of our planet's resources. But our ability to impart our genetic heritage to future generations is surprisingly vulnerable to any number of genetic, physiologic, or environmental foul-ups. Around one in ten couples suffers from some form of infertility.

In some societies, infertility is probably even more common. For example, in most parts of Africa, where infection of all kinds is so rampant, a huge proportion of women have damaged, scarred fallopian tubes. This can make conception impossible, because the egg cannot make progress down the fallopian tube and into the uterus where the embryo would normally implant.

In advanced Western society, infertility is common for a very different reason. More and more women in our sophisticated environment are developing careers and delaying pregnancy and child-rearing till later in life. But, as they get older, their ovaries are becoming depleted; as fewer and fewer eggs remain, infertility becomes an increasing likelihood.

Although the emotional burden of this problem often seems to lie more heavily on the woman and although certain types of infertility are increasingly prevalent, the physiological truth is that infertility is evenly balanced between the sexes. About a third of infertility problems stem from the woman's reproductive system, another third from the man's. The final third can be chalked up to either a combination of both partners or to reasons unknown and, in some cases, unknowable.

Human fertility is certainly a fickle attribute and the causes of infertility are very complex. A huge number of factors, it seems, may interfere with the process of making sperm. For example, while tight trousers, stress, and too much estrogen in drinking water have all been blamed for creating fluctuations in male fertility, this is a massive oversimplification. Until the mechanisms of spermatogenesis—the process by which the testis makes new sperm—are much better understood, it is likely that many of the key events responsible for male infertility will go unrecognized.

Color-enhanced X ray (hysterosalpingogram) from a young infertile woman. The womb cavity is the triangle in the center. The right tube is blocked near the ovary, and the left tube is blocked at the uterine end. She had two children after microsurgery.

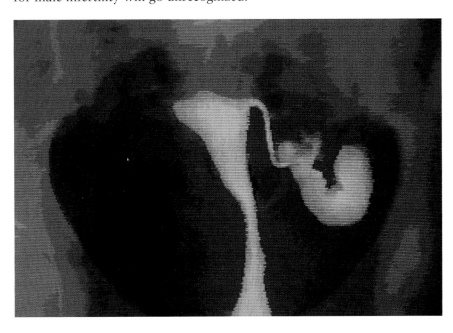

Certainly the testis and its delicate tubules can experience severe damage. In global terms, infections such as gonorrhea are probably the most important cause of tubular damage, but this only represents the tip of the iceberg. Most men who are infertile stop manufacturing sperm properly, or make sperm that are significantly abnormal and cannot function properly.

Why this is so is still a mystery. Possibly various environmental poisons or, more likely, those free radicals, which are associated with cancers (see Chapter 4), may sometimes be responsible. Probably, more often, there is a genetic cause. But to postulate a genetic cause represents a bit of an intellectual problem; if so much infertility is caused by some form of genetic failure, why didn't evolution take care of the problem? One would expect that infertile males would not pass on these genes and that they would die out in consequence.

> Human embryos are surprisingly fragile … A single embryo transferred to the uterus has only around a 20-percent chance of … becoming a baby.

Female infertility, too, may occasionally have a genetic basis. By far the most common cause of infertility in women is the failure to ovulate, or to produce eggs capable of becoming normal embryos. This is most frequently due to a condition called polycystic ovary syndrome. The classical description of women with polycystic ovaries shows that they are more prone than average to be overweight, frequently have skin problems such as acne, have mostly irregular periods, and sometimes are prone to abnormalities of sugar metabolism.

Not all women suffering from polycystic ovary syndrome are infertile. There are, however, good grounds for thinking that it has a genetic basis, as it often runs in families. The men in such families, though usually fertile, often show a degree of premature balding at the front of the scalp.

There are other potential genetic causes that may make women infertile. Sometimes the ovaries are prematurely depleted of eggs and ovulation stops when a woman is only 30 years old or even younger. Premature menopause such as this may well have a genetic basis; possibly at the time when the germ cells were streaming into the newly formed ovary in the developing female fetus, a genetic message went wrong.

We also have some recent evidence, from research in my unit at Hammersmith, that some women have a genetic problem that prevents implantation of their embryos. With the knowledge of genetics we are gaining, this should be curable.

Nevertheless, most female infertility is acquired. Worldwide, the most common cause is infection. Many different kinds of bacteria can infect the female genital tract, and, as a consequence, the tubes may become blocked, or the ovaries damaged and covered with adhesions. In such a situation, fertilization is unlikely, because the sperm cannot meet an egg. And, even on those occasions when an embryo is formed, it cannot pass down into the uterus.

In the developing world, where gonorrhea, tuberculosis, and various streptococcal infections are rife and often left untreated, infections are very common. And because pelvic inflammation frequently gives rise to very few symptoms—perhaps just transient discomfort and some vaginal discharge, for example—infection can go unnoticed. As a consequence, in Africa, the Caribbean, and many parts of Asia and South America, tubal damage is by far the most common cause of infertility.

Since the pioneering work of Patrick Steptoe and Robert Edwards in 1978, the treatment of infertility has become more sophisticated. Infertility specialists have had increasing success in helping us deliver our genetic missives to future generations. In vitro fertilization has changed our ability to help infertile couples and, although initially developed to treat women with blocked or damaged tubes, it is now applied to a much wider range of conditions. IVF has changed many aspects of human life. It gives us access to the earliest

Micromanipulation of egg and sperm, monitored on television. Direct injection of a single sperm into the egg has now become a commonplace procedure in the treatment of male infertility.

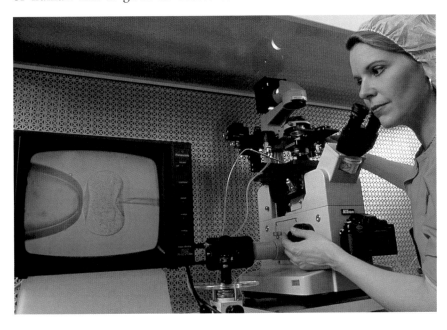

stages of development, to the human embryo. As we shall see, this opens up all kinds of opportunities that could be used or abused by society.

For IVF, drugs are usually given first to stimulate the ovaries to produce as many mature eggs as possible; these eggs are then removed from the ovary. Once harvested, they are mixed with a sample of sperm from the woman's partner, and, provided the mixture is kept under carefully controlled conditions that mimic those in the body, fertilization normally occurs.

Once an embryo has been formed, it can be transferred to the uterus during a very simple procedure. This is usually done about 48 to 72 hours after the eggs have been collected. At this stage, the embryo usually consists of about four to 16 cells and is totally invisible to the naked eye.

Human embryos are surprisingly fragile; that is to say, most of them perish for reasons that are still not entirely clear. A single embryo transferred to the uterus has only around a 20-percent chance of implanting, developing, and becoming a baby. In an attempt to beat these odds, two or three embryos are normally transferred. This increases the chance that at least one will survive, but it can also result in multiple birth. Sometimes, in countries where there are no regulations, or regulations are less strict, a greater number of embryos are placed in the uterus. This sometimes accounts for the headline-making births when five, six, or even more babies are delivered.

Multiple births are certainly not recommended; they can be dangerous, both for the mother and for the babies, whose chances of survival are tiny when they have to share a womb with five or six siblings. Even twins are at greater risk than average, and one in every 23 sets will die as a result. This is one reason why Britain decided to regulate this area of medicine. In the UK, regulations forbid the transfer of more than three embryos; and at my clinic in Hammersmith Hospital, we usually only transfer two embryos to avoid the serious risks caused by triplets. In the United States, implantation of more than two embryos is discouraged, but there is no law against implanting a greater number.

Although originally designed merely to bypass blocked fallopian tubes, IVF has turned out to be a huge breakthrough in treating male infertility. Using powerful microscopes and complex machines designed to undertake the finest manipulations in the laboratory, it is possible to pick up a single sperm and inject it directly into the egg. So-called intracytoplasmic sperm injection (ICSI) has been one of the big success stories in fertility treatment. But, as we shall see, this procedure, too, carries certain genetic implications.

All this sounds really promising, but does IVF deliver the goods in real terms? After 20 years of refinement and practice, IVF has not in truth realized the hopes of the early pioneers. The crunch comes both with its complexity and its expense. The likelihood of success with a single treatment is certainly not noticeably impressive. On average, in clinics across the United States and the UK, despite the care taken throughout the process and the expertise of the embryologists, only 15 percent of IVF treatments result in a live birth. The success rate in less proficient clinics is even lower. Considering the financial cost and the emotional stress involved, this is disappointing.

IVF is extremely demanding, both physically and emotionally, especially for the woman. Constant monitoring—daily hormone blood tests and regular ultrasounds—is necessary to make sure only mature eggs are extracted. The powerful drugs used to stimulate the ovaries can have unpleasant side effects, including enlargement of the ovaries, with abdominal pain and general sickness. Sometimes, these drugs can cause serious overstimulation, called "hyperstimulation syndrome," which can be disturbingly unpredictable and can cause sudden massive ovarian enlargement, internal bleeding, and shock. On rare occasions, it can even result in death. IVF, therefore, is certainly not to be considered a trivial treatment and is in real need of improvement.

The three stages of sperm injection into an egg. Remarkably, the degree of distortion caused by the injection needle does not seem to cause any lasting damage.

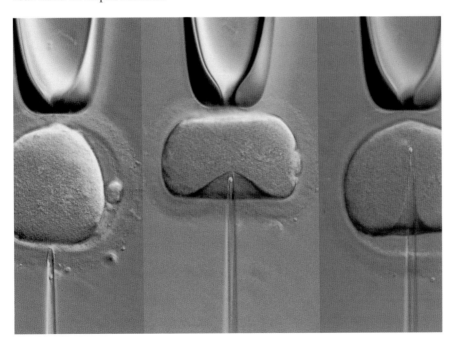

Controversies surrounding IVF persist. A recurring debate, as mentioned above, concerns the number of embryos that should be implanted. As women get older, they become much less fertile. Why this is so is not altogether clear, but, as women age, the eggs, which have been stored in the ovaries since before birth, gradually deteriorate. They also seem to develop more chances of being defective when they mature. Not only is a woman reaching her 40s more likely to be infertile, she is also much more likely to miscarry, because defective eggs lead to a defective pregnancy. By the time a woman reaches the age of 42, about 50 percent of her pregnancies will be likely to end in miscarriage.

For these reasons, many older women trying to conceive by IVF would prefer to have more embryos implanted and run the risk of a multiple pregnancy rather than have just one or two implanted and see the treatment fail. But the health risks to the mother, especially if she is older, are greatly increased if all the embryos survive. In a woman of 42, for example, quadruplet pregnancy will mean the high likelihood of a very complicated pregnancy, a very high miscarriage rate, and a huge risk of giving birth to babies that are too small to survive; if the babies do survive, they have a high risk of brain damage. The poor chances—that is, the failure rate—for older women may encourage some IVF centers, which want to boost their success rate, to take risks and give the patient too heavy a dose of drugs and implant too many embryos. If everything goes well, the center gets the credit, but it is the patient and her family, of course, who take the risk.

For couples who go through a number of cycles with no success, the experience can leave them feeling bitterly disappointed in IVF, but it is easy to be too negative. The development in IVF techniques has allowed many couples, who previously had no hope at all, a chance to conceive. Moreover, within a very few years, current research will make much of what we are doing at present seem extraordinarily crude. My own view is that IVF will get progressively easier and that fewer of the present-day damaging drug treatments will be necessary.

It is also important to recognize that the low percentage of IVF pregnancies cannot be mostly blamed on the technology. IVF cannot make good what is naturally bad. A substantial proportion of naturally produced human embryos have defects that do not allow them to survive the early stages of development. Between 25 and 30 percent of human embryos have abnormalities of their chromosomes. Others have cells that die prematurely or fragment for unknown reasons. Hence, even for a sexually active, fertile couple, pregnancy can often take months to occur.

Generally, if chromosomes are missing, or are simply abnormal, the genetic program for development cannot begin in earnest. There are exceptions: Down syndrome, for example, is caused by having three copies of a chromosome called trisomy 21. Although many of these babies survive pregnancy, many more do not. This failure of embryos with severe defects to survive acts as a kind of evolutionary filter to guard against large numbers of abnormal babies being born.

Much research has already allowed for improvement in nearly all the stages of IVF. One of the first was an improvement in the initial hormone treatment that is necessary to stimulate the ovary. Until recently, a very expensive hormone, called FSH, was extracted from the urine of menopausal women. After menopause, ovaries do not respond to normal levels of this hormone, so the pituitary gland, which produces it, responds by overcompensating—secreting far more than normal. This excess then ended up in the woman's urine, from which the FSH was chemically extracted and concentrated. This process, however, was lengthy and expensive, and a biologically derived hormone, which is a protein, can produce a side effect by stimulating the immune response of the body. In addition, there was the problem of biologically produced FSH varying in concentration.

A major step in modern therapeutics is to synthesize these naturally occurring hormones and genetically modify them to remove the flaws. In the case of FSH, cells are taken from the ovaries of Chinese hamsters and grown in culture until they multiply into a thick soup. But before they grow, they are injected with the human gene that produces FSH. The cells in the soup then produce large amounts of FSH, quickly and effectively, at a controllable concentration.

IVF is an expensive gamble that only the rich, or those who are lucky enough to secure experimental support, can afford.

Unfortunately, such genetically engineered drugs are not yet cheap. Tinkering with the genetic makeup of the cells will allow us to refine the FSH produced so that it no longer causes side effects and allows us to stimulate the ovaries more precisely and effectively.

A significant proportion of the costs involved in IVF are firstly, the hormone FSH, the cost increasing with the age of the woman (the older the woman, the more FSH she needs to stimulate the ovaries), and secondly, the time- and labor-intensive process of monitoring ovulation and removing the eggs. In the United States, the cost as a whole can be up to $7,800 per treatment cycle, and with only a one-in-five chance of a successful pregnancy, IVF is an expensive gamble that only the rich, or those who are lucky enough to secure experimental support, can afford.

We also need to be able to collect mature eggs far more easily than at present and with less cost. A solution that will revolutionize the whole of IVF treatment might exist. There is a supply of hundreds of thousands of tiny invisible immature eggs just waiting in the ovaries, a supply whose numbers dwindle rapidly with age. The surface of the ovary, called the cortex, is covered in minute follicles. This skin, just a few millimeters thick, is the storehouse for the eggs.

In a young woman, one cubic millimeter of cortical tissue, a piece the size of a pin-head, contains between 200 and 400 eggs. Because this tissue is on the outside of the ovary, we could very easily, probably with only a local anesthetic, collect a small portion of tissue using a hypodermic needle. Such tissue could then be collected and frozen in liquid nitrogen for potential use at a later time.

> The fabled biological clock *does* tick away – time waits for no man – and even less for women.

If we could somehow tap into this plentiful stash on a regular basis and if the eggs could somehow be persuaded to grow to maturity outside the body, then the entire IVF process would become much less privileged and would be accessible to rich and poor. It would be far cheaper because, instead of giving a huge dose of drugs to an average-sized patient, weighing, say, 130 pounds (60 kilograms), we would give the same drugs in minuscule amounts to a microscopic piece of ovary.

It would also be simpler and less stressful because, instead of monitoring the woman's progress by blood tests, ultrasound, and hospital visits every day, we would merely watch the tissue in the laboratory, a process that almost certainly could be automated. Moreover, the fact that there would be many more eggs to choose from at the time of fertilization would give us a far greater chance of producing a successful pregnancy. IVF, then, would be likely to be much more successful.

If we could mature eggs in this way, outside the body, there would be many other potential benefits too. It would give women much greater reproductive freedom. As mentioned before, the current trend in Western society is for women to develop a career, delay having a family, and seek a pregnancy at a later stage in life than ever before.

The age when women marry and the age when they have their first baby have increased steadily in most industrialized societies. But this increase in age brings increasing infertility. The fabled biological clock

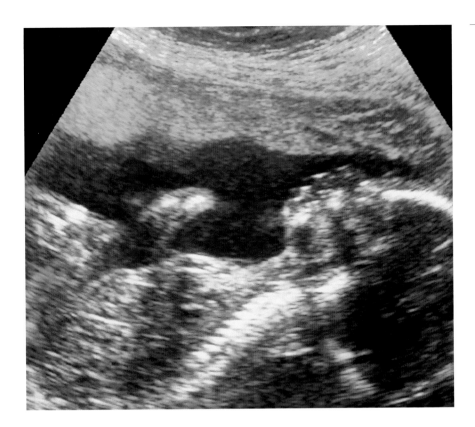

The baby halfway through pregnancy. Ultrasound scanning at 20 weeks can show remarkable details, including the baby's facial movements.

does tick away—time waits for no man, and even less for women. A woman of 42, wishing to become pregnant, probably has only one-tenth of the chance of success compared with a woman of 30. By the age of 45, the great majority of women—perhaps 60 or 70 percent—are hopelessly infertile even though they are still having periods and have not reached menopause. As we have seen, women in this age group are also more likely to miscarry and more likely to have a baby with a chromosome abnormality, such as Down syndrome.

Potentially, all these problems could be resolved with artificial egg maturation. Take, for example, the fact that there are very few female neurosurgeons; the field is very competitive, the training arduous. Egg storage and maturation outside the body could help more women become neurosurgeons, without jeopardizing the chance of having a family if they wanted. A young woman about to train to be a doctor could, thanks to a ten-minute minor surgical procedure, have her ovarian tissue stored before she enters medical school at the age of 21 or 22. She could receive her medical degree in her mid-20s, then train as a surgeon and become licensed to practice surgery in her early 30s, go on to specialize in brain surgery and do internationally recognized research, and at 40 become a leading consultant. Then, married or not, she could start to plan a family, knowing that she will be as fertile as she was at the age of 21 because her stored eggs are as pristine as those of a much younger woman.

Harvesting ovarian tissue has other, more realistic advantages. Quite a substantial number of young women and adolescents develop cancers or leukemia. Modern medicine, with all its improvements, has increased their chances of survival by curing many of these malignancies occurring in younger people. For example, the prognosis for leukemia has improved greatly in the last 20 years or so. But treatments for most of these malignancies involve cytotoxic drugs—chemotherapy—or radiation.

These treatments produce their effects by killing cells, and the egg cells in the ovary, like the cancer cells, are very sensitive to such agents. Cancer treatments increase the likelihood of early menopause; even when there is no menopause, sterility is a frequent outcome. Storage of ovarian tissue *before* cancer treatment has the potential for alleviating one of the worst aspects of cancer therapy—infertility.

Even without cancer treatment, premature menopause is quite a common problem. The cause of spontaneous menopause usually is not known; it is thought that it may be genetically determined in some women. Such women have no chance at all of becoming pregnant. Because there are far fewer babies available for adoption than people wanting to adopt, women with premature menopause, which can occur as early as the age of 20, are condemned to childlessness. Their only realistic alternative is to have a donor egg fertilized outside the body with their partner's sperm and then transferred to the uterus.

Of all the forms of IVF treatment, donor egg treatment is highly successful and the most likely to produce a pregnancy at the first shot; about 40 percent of egg donation treatments are successful the first time in young menopausal women. The problem is that it is extraordinarily difficult to find egg donors. And, currently, when we do, the donors must have the full panoply of IVF if their eggs are to mature after receiving large injections of FSH. They need intensive monitoring, ultrasound, and blood tests on a daily basis. It is scarcely surprising under the circumstances that only the most altruistic of donors are prepared to go through this extensive process. However, if there was a resource of frozen ovarian tissue banked by women who had completed their childbearing, donor egg treatments would become simple and commonplace.

A new technique, such as egg maturation outside the body, naturally comes with a checklist of troublesome hurdles to overcome. The first problem is thawing the tissue without damaging the eggs. Then the individual follicles, each containing one egg, have to be separated from the tissue. These two steps are relatively easy compared to the next problem, which is enticing the eggs to mature, in vitro, to the stage when they can be fertilized.

The culture in which we grow these eggs is critical, and no one is entirely sure which hormones and growth factors are necessary to complete the maturation. The initial growth stage has been completed, using tiny amounts of the hormone FSH and various other factors. The task now is to find the exact conditions and genetic messages to mimic, which allows the eggs to mature and stay strong and healthy, ready for fertilization.

There are other reproductive techniques being explored to overcome the problems that women face as they grow older. One, which is still at an early stage, is that of cytoplasmic transfer. Just exactly why the eggs of older women are less fertile and more likely to be defective is unknown. It is all tied up with the aging process, which is thought in part to be due to mutations in that tiny part of the DNA that is held in the mitochondria.

Mitochondrial DNA, which is responsible for controlling our energy usage, does undergo change with age; for this reason, some scientists think that mitochondria transferred from the donor eggs of younger women might be a way around this problem. Perhaps the mitochondrial-rich cytoplasm fluid is a kind of royal jelly, an elixir of youth. Various strategies are being developed at present. One treatment involves taking out around five percent of the cytoplasm but none of the nuclear DNA from the egg of a healthy, young donor, and injecting it into the eggs collected from the older patient. That egg, strengthened by this cytoplasmic transfusion, is then fertilized with just one sperm using micromanipulation.

Jane and Manfred Steiman went through repeated IVF, and Jane's story is a perfect example of how modern infertility treatment seems to conflict with the notion of the Superhuman.

Manfred runs a successful small business involving computing; Jane teaches German part-time. They married 17 years ago and live in Hertfordshire, England.

After seven years of trying for a baby and seeing various doctors in different clinics, Jane had two cycles of IVF costing in total around £5000 (about $7,500), but she didn't conceive.

The drugs made her bloated and gave her a recurring headache.

"But the worst thing," says Jane, "was waiting 12 days after the embryo transfer to see if I was pregnant. I didn't feel I could do anything, not even the dishes, in case I dislodged a developing baby. And I had morning sickness, even though I knew that it was far too early to be *really* pregnant. But worst was when my period came—a day or two late."

For the Steimans, going through IVF, and Jane's reaction to it, had a pretty bad effect on their relationship and sex became pretty infrequent. Always regular in her menstrual cycle, five months after the second IVF failed, Jane missed a period.

"Without telling Mannie, I slunk into the local pharmacy and bought a pregnancy test kit. The following morning, after he had gone to work early, I trembled as I took it out of its package. When the test was positive, I was sure I was hallucinating. It just seemed impossible to have fallen when I was hardly doing anything active about things."

Jane's doctor confirmed the result. Manfred was overjoyed and, six days later, they went together for an early ultrasound scan at the clinic. Jane remembers that she had never felt as close to Manfred as she did when they first saw the baby's heart beating on the scan. Later that day, they made reservations for a celebratory dinner at a local Italian restaurant.

About six o'clock that evening, Jane went to the lavatory to urinate. When she looked into the toilet bowl, she felt faint. The water was deeply blood stained; she was bleeding.

Manfred was home within 20 minutes of Jane's call, but the wait in the emergency room seemed to go on for ever. After Jane lay on a bed for three hours, The Steimans saw an obviously harassed young doctor who clearly thought that a possible miscarriage was not all that much to get excited about. There were more important emergencies elsewhere. At 11:30 PM, Jane was sent home to rest in bed.

Four days later, the bleeding had turned brownish. But another scan showed no sign of a heartbeat.

"The worst thing," says Jane, "was that nobody seemed to think that this shattering event was particularly important. I felt numb; I couldn't look at Mannie. One nurse said, "Don't worry, dear, you've done it once, you can always do it again." One of the doctors said, "The baby was probably abnormal—the miscarriage was nature's safety valve." However, I couldn't get it out of my mind that I had done something to cause the abnormality, or that the whole thing was some kind of punishment."

In the next five years, Jane repeated IVF, with three embryos being transferred each time. Six further treatments and £20,000 (about $30,000) later, Jane and Mannie couldn't believe that they had finally lucked out. Within three weeks of the embryo transfer, Jane felt very sick, and ten days after that a scan showed not one heartbeat, but three. Jane was carrying triplets.

Over the next six weeks, the Steimans had difficulty in keeping the wonderful news to themselves.

"Having been shattered before by a miscarriage, I wanted to wait before telling my mother, or my closest friends," says Jane. "Last time, I didn't know how to look at anybody afterward and started to weep if I met friends with small children or those who were pregnant. And everybody seemed to be getting pregnant."

Just ten weeks after the embryo transfer, Jane started to feel pains in her tummy, very low down. Manfred called the doctor and she was rushed to the hospital. Just 24 hours later, Jane needed an anesthetic to remove the bits of blood clot, membrane, and placenta following a miscarriage of all three babies.

It took a long time for her to recover. For several weeks she couldn't work and her sex life stopped completely.

"I felt like an empty vessel," says Jane, "but I also knew I couldn't face IVF again."

After two years of intermittent depression, a trial separation from Manfred, loneliness, and then reunion, Jane read in a woman's magazine that there were alternatives to IVF. She turned up in our clinic at Hammersmith, which is where she and I first met. Four months later, after a simple operation to clear her fallopian tubes, Jane was pregnant naturally. Rupert was born when Jane was 41 years old, healthy and well.

"I realize now that it was better for my body to be coaxed into working than to be forced. And IVF seemed like brute force. . . . "

If IVF fuels controversy among certain sections of the population, then some current and future developments in genetic diagnosis and manipulation seem to make these people apoplectic. Doctors are thought to be taking control of reproduction like never before; the potential for genetic manipulation is at the heart of the ethical dilemmas and the accusations that fertility doctors are playing God.

This is far from true; all the evidence suggests that our knowledge is being used and will continue to be used responsibly. There are immediate and valuable benefits, which it would be senseless to refuse. Helping infertile couples is valuable to them and to society; it is one of the best ways we can promote the much-vaunted family values. There is nothing wrong with helping older women have a child, provided they can give their offspring the care and nurture that is essential to all good upbringing. Indeed, by helping these women, we are augmenting and nourishing their life experience and enriching society. There should be no serious objection to our attempts to prevent miscarriage or avoid serious genetic diseases, which cause such torture to affected children and pain to their families.

But having said that, IVF techniques do present a challenge if not a threat, as well as a promise. The "designer" baby, the baby who will have what are perceived to be "desirable" attributes, undoubtedly represents a threat to human relationships and family values. And the most basic form of this kind of intervention is our ability to select the sex of our babies.

> The "designer" baby … undoubtedly represents a threat to human relationships and family values. And the most basic form of this kind of intervention is our ability to select the sex of our babies.

This desire is hardly new. It is something that humankind has wanted to do, and tried to do, for thousands of years, mainly because, in many societies, male babies are preferred to females.

Countless methods have been proposed to help load the dice in determining a baby's sex. The ancient Greeks, including Hippocrates, thought that boys came from a fusion of "humors," a fusion of the four basic human temperaments. Boys came from the right testicle and girls from the left. They thought, therefore, that the position in which a couple had sex would favor either a boy or a girl. Anaxagoras, a contemporary of Hippocrates, suggested that the left testicle should be tied off with string to ensure the production of a boy. This advice was taken to an extreme by some French noblemen in the 17th century, who underwent castration of the left testicle in order to try and produce a male heir.

More modern, but no more effective, is the practice of vaginal douching. Since the 1930s doctors have claimed that sperm with a Y chromosome, which would produce a boy, prefer an alkaline environment, and the sperm with a X chromosome, which would produce a girl, prefer an acid environment. This theory is still alive and well despite a complete lack of evidence to support it. Proponents of this theory say that women should douche in a solution of bicarbonate of soda before sex if they want a boy and in a diluted solution of vinegar if they want a girl.

Currently, some clinics offer sex selection by using a sperm separation technique. This depends on the fact that X and Y chromosomes are different in size; the Y chromosome is very slightly smaller, and there is, therefore, a tiny difference in the weight of X and Y sperm.

One version of sperm separation involves spinning the sperm in a centrifuge. The idea is that the lighter Y-bearing sperm are more likely to be thrown out to the very edge of the centrifuge. Another technique involves allowing sperm to pass through a fluid of variable density. The idea here is that "heavy" sperm will be trapped or retarded and "lighter" sperm advanced. At least one clinic in London has offered this technique, with apparently no shortage of customers; most of them, rather predictably, wanted a boy rather than a girl.

However, there is very little independent evidence to back up this procedure's claims to success. It is not clear whether the almost imperceptible weight difference of the sperm is anywhere near significant enough to make a difference in a centrifuge, in fluids of variable density, or with these two techniques combined. In my view, patients who attend these clinics are not getting value for their money, although the couples who do conceive a baby of the required sex—I imagine these make up no more than half the clients—may be deluded into thinking the method actually works.

There is one completely reliable method for selecting the sex of your baby. At Hammersmith Hospital we have developed a technique called Preimplantation Genetic Diagnosis (PGD), which can be used in conjunction with IVF. This technique allows us not only to determine the sex of an embryo, but also to detect the presence of genes that would lead to severe hereditary conditions such as cystic fibrosis and beta-thalassemia.

We feel strongly that this invasive technology should be used only when there is a real risk of genetic disorder—for example, when families carry genes that might cause the death of a child if he is a boy rather than a girl. This stance has been ratified by the Human Fertilization and Embryology Authority; under its rules, PGD can be used for sex selection in the UK only if there is a chance of a sex-linked hereditary disorder.

There are hereditary disorders that are carried by females but only affect males. For example, hemophilia, which afflicted the Russian royal family and was also carried by Queen Victoria, suppresses blood clotting and can lead to uncontrolled bleeding. Some children can survive this condition perfectly well with occasional blood transfusions and regular doses of the proteins that are important in blood coagulation. Others have a terrible time. The slightest injury has catastrophic effects and, for this reason, they can never be allowed to play normally or undertake activities the rest of us take for granted. Even then, sudden massive spontaneous hemorrhages can be life-threatening.

The most important of the sex-linked conditions is Duchenne muscular dystrophy, simply because it is the most common. In the United

States around one boy in 4,000 suffers from this condition. It causes progressive muscular weakness; these children are generally confined to wheelchairs, and can live only while the muscles that allow them to breathe are strong enough.

It is very unlikely that PGD, which involves IVF, will be the preferred method of sex selection in the future. It is too complicated. Sperm separation, possibly using machines that can detect male or female DNA, will almost certainly be perfected for human use. This method is already possible in farm animals. It could become relatively cheap, cheaper than IVF, and only requires artificial insemination as a medical intervention. Once such techniques of sex selection are simplified, the process will become even cheaper and more accessible. This path, however, certainly has its dangers. We might find that the practice of sex selection is increasingly difficult to police and, like abortion before it became legal, black-market sex-selection clinics could be a lucrative business.

In the Western world some religious groups have a cultural preference for boy children, a preference that for them has social and financial considerations. Curiously, there is also one section of the British population that has shown a deep interest in it. During debates in the British Parliament on the morality of modern reproductive treatments, a desire for the ability to select the sex of a child was expressed by a number of hereditary peers. Perhaps, given their dwindling numbers, they were worried about producing a male heir; indeed, during a 1990 debate on human embryology in the House of Lords, some peers seemed much more interested in the question of inheritance and the succession of their own peerage than the actual subject of the debate, which concerned the use of donor sperm in artificial insemination. They did not, however, garner a great deal of sympathy. Although I visit the House of Lords on a regular basis, I have yet to meet any English hereditary peer who, emulating the French noblemen, has gone to the extremes of cutting off his left testicle.

Developing nations such as India and China may in the future have a more serious problem on their hands. It is easy to imagine that in these countries there will be a great demand for clinics that offer a quick, cheap, and effective method of sex selection. This carries obvious dangers. Some people believe that changing the ratio of boys to girls could have very undesirable social side effects, leading to an imbalance in the population. Another argument against sex selection points out that these treatments will be available only to those who can afford it, thus making social and economic inequalities even more pronounced.

Once an egg (no matter who produced it) has met with a sperm (under whatever circumstances) and been implanted in a uterus (in whoever's body), there are still nine difficult months ahead. About 15 to 20 percent of all pregnancies end in miscarriage, most in the first few weeks after conception.

Even then, once the hurdles of conception, pregnancy, and childbirth have been surmounted, we cannot breathe a huge sigh of relief. The World Health Organization estimates that nearly 2.2 million infants died in 1998 in its member states. The US March of Dimes points out that one in every 28 babies in the US is born with a birth defect, which ranges from heart defects and metabolic disorders, such as Tay-Sachs disease, to fetal alcohol syndrome.

And, according to the US Centers for Disease Control and Prevention, those birth defects are responsible for some 20 percent of all infant deaths. In Britain, the situation is essentially similar. Birth defects, together with premature delivery, are now the most common cause of babies dying, and the incidence is slightly higher in poorer communities where prenatal screening is less effective.

There are, of course, monetary values associated with these statistics. In 1992, caring for children with birth defects cost the American people more than $8 billion. There are no comparable statistics available for Europe, but the costs of care are likely to be broadly similar. In Cyprus, 40 percent of the health budget goes to treating children with beta-thalassemia. Mediterranean countries have a high proportion of their population carrying this gene defect, and worldwide it causes the death of over 250,000 children annually.

Beta-thalassemia is an inherited condition that prevents the proper formation of hemoglobin, resulting in a severe, sometimes fatal, anemia. But percentages and pounds cannot begin to describe the psychological impact of the loss of a newborn child, or having to deal with the emotional consequences of raising a significantly disabled child, or, indeed, the suffering of children who have to live with these diseases.

Fiona and Peter Hingston just *knew* there was something wrong with their newborn daughter. Her diapers were stained green and she kept throwing up. But after observing her in hospital for a weekend, her doctors could not see anything wrong; she was just colicky, they said. Green diapers are perfectly normal for a breast-fed baby. "Take your baby home and don't worry," was the advice the Hingstons were given. And so they did. But Fiona could not stop worrying.

Jade did not seem to be getting any better. Fiona watched as two of her friends' babies put on weight each week. But in that time, Jade had only put on half the amount they gained. Then Jade's weight gain suddenly dropped to about two ounces a week, while the other babies were still putting on pounds. She was still throwing up after every feeding, and her diapers were still green. Then she started to get very pale. Finally, Fiona and Peter knew for sure that something was wrong.

When Jade was around nine weeks old, Fiona and Peter found a doctor who listened to them. Fiona took Jade from their home in Torquay, England, to Treliske Hospital in Penzance. By that time, Jade was looking so pallid that her appearance shocked even the nurses. Three weeks, two transfusions, and a good many medical tests later, the doctors were still at a loss. It was then they belatedly decided to do a sweat test.

The sweat test is the gold standard for the diagnosis of cystic fibrosis. A chemical is applied to the skin to cause sweating and the sweat is then analyzed to determine how much chloride it contains. When chloride levels are high, they indicate cystic fibrosis. Jade's chloride levels came back high.

Cystic fibrosis (CF) is the result of a gene mutation that affects the way in which salts and water are passed through the cell membranes. This produces a chain reaction of side effects, but the main problem is that the respiratory system gets clogged up. Chronic infections take hold and eventually cause even more lung damage. This is why cystic fibrosis is potentially so serious.

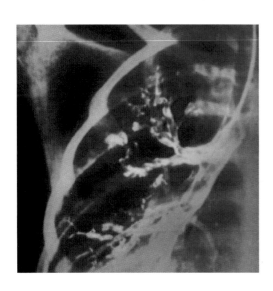

ABOVE *The chest X ray of a boy with cystic fibrosis. Many of the tubes in the lungs, the bronchi, are blocked with thick mucus (colored green), making breathing very difficult.*
OPPOSITE *A close-up view of the lung lining in cystic fibrosis. The hairs on the surface of the cells, called cilia, normally keep the lung passages clean. When the mucus builds up, their job becomes impossible. This is one of the most common genetic defects in Britain.*

"I remember the doctor coming in," says Peter. "We were by the crib in the hospital, and when they said "CF," all I knew was that they died at an early age. I thought, 'Well, she's going to die,' and that was that. I just broke down and cried. For a couple of days I just couldn't handle it."

Peter and Fiona pulled themselves together again and started listening to the doctors. What they heard was relatively encouraging. Cystic fibrosis patients born in the 1990s can expect to live for around 40 years, if they are lucky. And because Jade's cystic fibrosis seemed to be mainly concentrated in the tissues of her digestive system rather than in her lungs, she has not gotten the

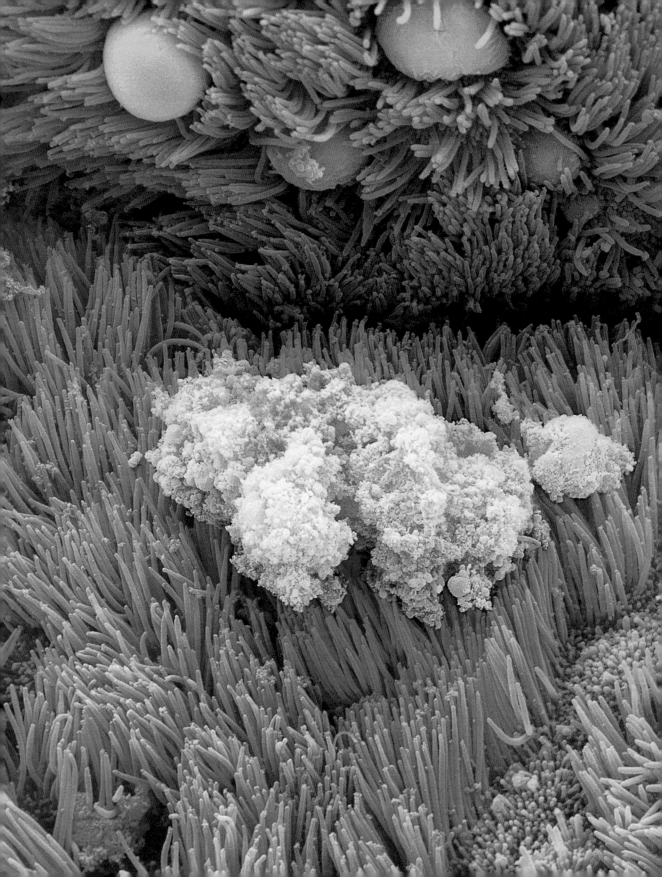

constant respiratory infections that can plague most other cystic fibrosis patients. She takes enzymes to replace those not produced in her digestive system, which allow her to process her food and gain weight. Twice a day, Fiona and Peter have to pound on her chest and back to break up the thick, sticky mucus that threatens to clog up her other organs.

In the UK today, there are more than 7,000 people living with cystic fibrosis; and one in every 2,000 children has the disorder. It is the country's most common, life-threatening, inherited disease. Cystic fibrosis is a recessive genetic defect. This means that if a person carries just one defective gene he or she will not have the disease, but if two such people have a child, there is a one in four chance that their child will be born with cystic fibrosis. In the UK, one-in-20 adults carries the mutated gene.

Like most other people in this situation, Fiona and Peter had no idea they were cystic fibrosis carriers, but they now know that any child they might have together in the future has a one-in-four chance of being born with cystic fibrosis. They had always wanted a large family and couldn't imagine Jade as an only child, but they could not take the risk of having another child with cystic fibrosis. So, Peter and Fiona decided to try preimplantation genetic diagnosis.

The first three couples we treated with PGD, in 1990, already had a child who had died of a sex-linked genetic disorder consisting of three

Cystic fibrosis patients require repeated physicaltherapy each day to encourage good drainage of their lungs. This can be a quite a difficult process when dealing with a boisterous or cranky child.

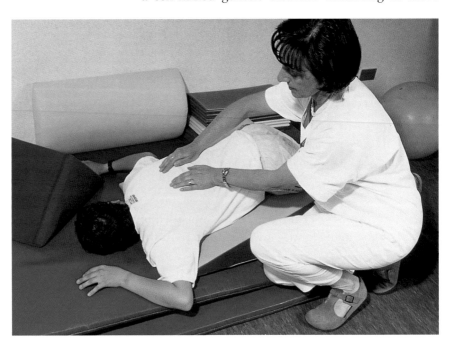

terrible and sometimes fatal diseases: muscular dystrophy, adreno-leukodystrophy, and severe X-linked mental with physical retardation.

The first parents we treated knew that if they had another boy there was a high chance of the same terrible sex-linked disease striking again. In these instances, the PGD was designed to select only female embryos because, even though they may be carriers, girls do not suffer from these diseases. We collected and fertilized the mothers' eggs in exactly the same way as for standard IVF treatment. After three days, when the embryos comprised just eight cells, a single cell was removed from each embryo. This process is extremely delicate and difficult to do without damaging the embryos. Indeed, it was such a pioneering technique at that stage that most doctors thought our approach would be impossible.

Once we captured the individual embryonic cells, they were immediately frozen and, a minute or so later, thawed to release the DNA contained in the nuclei. The DNA was analyzed to determine whether or not the embryos were male or female. We then picked out two of the female embryos and implanted them. Two of the women became pregnant, and both gave birth to twin girls. These successes opened up a world of possibilities.

By 1992, the technology of DNA analysis had become more sophisticated and we were able to make a specific diagnosis of a particular gene defect. It is not surprising that we chose cystic fibrosis, because it is the most common fatal genetic disorder. We could tell, reliably, which embryos would develop to have the condition, which ones would be carriers (that is, pass on the mutated genes but not succumb to the condition themselves), and which embryos were completely free of the cystic fibrosis mutation.

By these methods, it is possible to test for these conditions once a woman has become pregnant; using the same kind of genetic analysis we are able to diagnose the condition in an early-stage embryo. The parents we treated at Hammersmith came for the highest moral reasons. Firstly, they did not want any child to suffer from a life marred by pain and followed by premature death. Secondly, they were extremely reluctant to have abortions if prenatal tests turned out to be positive. They would much prefer to have healthy embryos implanted in the first place, despite the rigors and uncertainties involved in IVF.

PGD can now detect some 30 or so different genetic disorders, and, with better knowledge of our DNA, the number of diseases that we can detect grows each month. Once the Human Genome Project is completed, and the genes in the genome identified, this work has potential for great expansion. It is now possible to examine chromosomal as well as genetic disorders. Using bright fluorescent dyes, researchers have learned how to stain the DNA in different chromosomes different colors.

A defect of the chromosomes, rather than individual genes, is the most common cause of miscarriage. In fact, around 20 to 30 percent of human embryos have one or more abnormal chromosomes. A few chromosomal defects allow the embryo to survive, but with severe consequences, as, for example, in Down syndrome, in which there are three copies of chromosome 21 instead of two. These children are born severely handicapped, with profound mental retardation, and often have other problems, such as an abnormal heart.

The chromosome staining or painting method can be used to detect which embryos only three days old carry these defects. The method can be used for women who have suffered repeated miscarriages, or repeatedly failed IVF, since the cause is often a particular kind of chromosomal abnormality. As already mentioned, older women are especially prone to these problems, and it is hoped that the success rate of IVF can be improved dramatically using these PGD techniques rather than simply implanting the embryos and hoping for the best.

Inevitably, the procedure is fraught with difficulties. One, which has largely been overcome, is the microsurgery necessary to remove an embryonic cell. A three-day-old embryo is extremely delicate. It is not easy to remove a cell without damaging the embryo, which would obviously prevent any further development. PGD in general is also very time-consuming, and labor intensive, and so it is, therefore, expensive. It is unlikely that the process can be automated in the near future, and for the moment there is little chance of PGD becoming a routine procedure that would allow for the screening of large numbers of human embryos. This is all on top of the uncertainties inherent in IVF itself.

In the longer term, PGD may become much simpler, as well as more sophisticated. Efforts are now being put into examining embryonic cells using a method known as array technology. With this approach, a whole variety of genetic predispositions to disease might be screened rapidly and automatically, with a diagnosis available in a few hours. Once more is understood about the human genome, genetic screening may become more easily available on a wider basis.

Unsurprisingly, PGD has met with a great deal of controversy. The whole principle of screening embryos, which inevitably means choosing some and discarding others, is unacceptable for Catholics and others who believe that life starts at the moment of conception. Most people, however, believe that life grows in importance with development. They believe that a clump of cells is very far from being identifiably human, and that at this very early stage, when the cells have not yet been organized into the shape of a body, we do not need to grant them the full rights of a human being. For the many people who disapprove of abortion, PGD is a happy medium because healthy embryos can be selected when they are just three days old.

Some advocates for the disabled, and some disabled people themselves, fear that PGD is the beginning of genetic cleansing. They point out, quite rightly, that many people with all kinds of congenital conditions live healthy and happy lives. And, in their view, PGD is dangerous because it devalues the status of handicapped people, reduces their dignity, and increases prejudice toward them.

> Some advocates for the disabled, and some disabled people themselves, fear that PGD is the beginning of genetic cleansing.

However, I do not think that this is a correct analysis. From many meetings with handicapped people, mostly concerned with their anxieties about passing on their disorder to their children, I have formed the strongest impression that they believe this is a matter that must be left to the individuals most afflicted by these disorders. I disagree that the use of this technology in any way changes the status of handicapped people. A person, once born, has inalienable human rights and must be accorded immutable human values.

British polls have repeatedly shown that most people do not object to the use of these procedures for conditions such as cystic fibrosis or muscular dystrophy. This is because they severely affect the quality of life and ultimately prove to be fatal. However, there is a body of opinion that considers that some nonfatal conditions, such as restricted growth or congenital deafness, might be included in the list of genetic defects for which we might wish to screen. Advocates take the position that the solution to the problem of disability lies in dealing with the problem of discrimination and prejudice, rather than in removing disabled people from society.

They have a valid point. Such arguments are founded on real fears, not only of genetic cleansing, but also of the creation of a genetic upper class. After all, given that it is unlikely in the foreseeable future that nations will use their public sector finances to fund techniques such as PGD, this could pave the way for a private sector boom in designer babies, available only to the rich. Why stop at deafness or other conditions that seriously affect the life of the child? Why not select for intelligence, good looks, strength, or blonde hair? Why not pick out the best-looking, most aggressive, alpha embryo in the petri dish?

Practitioners in this field in Britain, and the politicians who make the laws, are not complacent about these dangers. Even if we knew which genes would guarantee an IQ of 150, or which genes would predispose us to high cheekbones and blonde hair, we could not use this information because there is a clear-cut ban on using PGD for anything other than serious congenital defects. The position, however, is not so clear cut in other countries, including the United States. And we, as a global society, now need to decide what constitutes a serious condition and to what extent we should implement these most challenging techniques.

Fears have also been expressed about screening for genetic disorders that occur later in life. It is now possible to test embryos for some inherited cancer genes because certain embryos are strongly predisposed to various kinds of cancer: familial breast and ovarian malignancy, polyposis coli (which causes bowel cancer), and some forms of brain cancer. Someone who carries one of these genes has a high risk of developing cancer at a young age, and, because of his or her genetic predisposition, these cancers can be very hard to treat; they spread quickly and secondary tumors can spring up more easily.

Such people can, of course, lead full productive lives until the cancer strikes. Therefore, the question is this: Should we discard embryos that have a potential life of 30, 40, or even 50 years, during which time they could be completely free from any symptoms caused by the fatal genetic disease that they carry?

Twenty years ago, in Cyprus and in Sardinia, one in seven children was born with a crippling genetic disease, beta-thalassemia. This is a condition that affects the production of hemoglobin in the red blood cells. If a baby inherits the beta-thalassemia gene from both mother and father, by six months old it will suffer from severe anemia. Thereafter it may require a blood transfusion every one or two months to keep it alive. The tissues that make red blood cells – the bone marrow and the spleen inside the abdomen – become hyperactive to compensate for the anemia. These children, and young adults if they survive, become susceptible to infections and often develop heart failure, liver failure, or diabetes. This is because their internal organs become loaded with iron that cannot be properly utilized in normal blood cell production. The extra iron load causes organ failure and most of these young people are dead by their early twenties. On the exceptional occasions when a bone marrow transplant that matches their tissue type is available, there is a chance that the condition will be cured permanently. But this is a rare gift, and most sufferers just see themselves gradually deteriorate.

The condition is recessive; a person can simply be a carrier of the abnormal gene from one parent, in which case they simply suffer mild anemia, which causes few symptoms. However, if both parents are carriers there is a one-in-four chance that any child they conceive together will suffer from the severe form of the disease; two out of four children would be carriers and likely to have mild anemia; the chance of the child being entirely free of the gene is just one in four. So, even though most children who have the condition do not survive to have children of their own, the number of carriers in the population ensures that the disease survives through the generations. Beta-thalassemia is something that Sardinia has had to learn to live with.

One of the gene mutations for beta-thalassemia has probably been in the Sardinian population since the time of the Stone Age. Bones over 4,000 years old in their archaeological sites are now being excavated. These show deformities consistent with the damage associated with thalassemia. These deformities are caused by the overactive bone marrow. Scientists are trying to analyze DNA from these ancient human remains to see if mutations that cause this disease were present long ago.

There is also compelling evidence from another source. Sardinia is a mountainous, barren island, whose sparse population has low fertility. The Sardinian people are so isolated that they still speak a version of Latin, rather than the more modern Italian dialect. It is a country known for its colorful costumes, its peasant music, and dances; some dances can be traced back to early history.

Each June, Sardinians hold festival dating from ancient times that seems to celebrate a fertility rite. The men dress up in goatskins with cowbells on their backs and dance around women whom they attempt to lasso. This so-called demon dance is interesting because the participants wear masks of a remarkable kind—some made of terracotta, others of wood. Many are very old and have been in the families for generations. But the really curious thing is that many of these masks show an extraordinary disfigurement—they have high, prominent foreheads and sunken faces. This is the characteristic stigma that beta-thalassemia leaves on the face of its victim.

All over the Mediterranean, there has been a substantial drop in the number of babies born with the condition. A DNA test can determine very early in a pregnancy whether the fetus is carrying the prevalent form

The Mamathones of Sardinia. Their ancient masks indicate the features seen in advanced beta-thalassemia. It is clear that this disease has been prevalent on the island since prehistoric times.

of the beta-thalassemia gene. Hence the mother can abort the fetus at this early stage if necessary. In Cyprus, widespread testing, together with arranged marriage and prenatal diagnosis, has managed to control this terrible disease. Although this is not a complete solution, it has certainly reduced its incidence.

Neither PGD nor abortion are ideal long-term solutions to the problem of beta-thalassemia. Even if they were undertaken for every conception in Cyprus, and the affected children thus prevented, these recessive genes would continue to be passed down by carriers through the generations. Moreover, PGD is especially demanding, expensive, and complex, so much so that even routine IVF is unavailable in Sardinia and difficult to get in Cyprus. So what can we do to win the war against recessive genetic disorders such as beta-thalassemia? Is there a better solution?

Possibly there is. We could go one big step further. Rather than picking and choosing human embryos, as we do in PGD, we could actually introduce new genes during human development. These genes could replace or mend those faulty or missing genes that cause or carry these disabling and fatal congenital diseases. It is a strategy that could in theory reduce the incidence of conditions such as betathalassemia and cystic fibrosis in the world to a much greater extent.

Such a process would involve introducing a transgene, or "foreign" gene. Transgenic technology is one of the most fundamental and remarkable advances in human medicine. Our ability to make transgenic animals, mostly mice, has enabled scientists to study how genes work, what happens when a particular gene is missing or modified, and the effects produced when different genes interact. Transgenic mice are also a very important model for human disease processes and have been of immense importance in our understanding of cancer, infection, diseases affecting brain function, and so on. But transgenic technology, of which cloning is one small area, carries with it difficult and deep-rooted ethical questions.

With the exception of cloning techniques, most new genes are introduced just after the eggs are fertilized. Embryos are collected before they have divided into two cells and a new gene sequence is injected into the nucleus of the cell. If the sequence becomes integrated and starts to work normally, it becomes an integral part of the genetic makeup of the embryo. Thereafter, the new gene sequence will find its way into every cell in the body. It is just possible that missing or defective genes, which cause conditions like beta-thalassemia, could theoretically be replaced by normal benign genes that remove any risk of the condition taking hold in later life.

Transgenic technology really began life in 1980 when Dr. Jon Gordon and Dr. Ralph Brinster, among others, injected pieces of new DNA into mouse eggs. Some of the first experiments actually involved injecting a mouse egg with a human gene, a gene that helps make hemoglobin in red blood cells. One team was using human growth hormone genes to make "giant" mice. These, rather sadly, were not huge three-foot mice, but just mice who grew up a little faster and became slightly bigger than normal.

At first it might seem incredible that human genes can find a home in the genome of an entirely different species, such as mice. But the molecule that makes DNA is essentially the same molecule in all animals, indeed in all living things. Even the sequence of base-pairs, the letters of the DNA code that make up the molecule, is incredibly close to that of all mammals. There is a small percentage of difference overall. There are great similarities in the genes that control basic functions such as making blood cells. Because these are well conserved throughout evolution, human DNA can often fit in extremely well and function normally in an animal such as a laboratory mouse.

It may seem even more remarkable that transgenes from mice can be injected into the embryos of fruit flies. Slotted into the gene cascades in fly DNA, the mouse genes may actually function in a species that has been evolutionarily separate from mammals for over 500 million years. One experiment involved a gene from a mouse that controls the development of the eye. This gene, Pax6, was injected into the leg of a fly embryo. Pax6 is a so-called "master gene,"—it sits at the top of the genetic chain of command. Having initiated the set of instructions that says "make an eye," incredibly, as the fly embryo developed, an eye appeared on its leg. However, the eye was not a mouse eye—it was a perfectly formed insect eye. Pax6 from the mouse had told the fly genes to make an eye and they obeyed, carrying out the instructions according to the fly's genetic code.

This commonality between species is the reason that we may be able to breed transgenic pigs with organs that are suitable for transplantation into humans (see Chapter 2). The genes that make the "human" proteins, which cover the surface of the organs (and hence look friendly to our immune systems), are readily accepted into the pig's DNA and take their place in the genetic program without the whole system breaking down.

Getting transgenes to work is not easy. Usually a number of copies of the gene, or a particular sequence within a gene, is injected into the fertilized egg in a solution. Some experiments involve

A baby mouse carrying the transgene that expresses green fluorescent protein. This harmless green pigment, derived from the jellyfish, is widely used to demonstrate the presence of genetic lines that have been artificially introduced.

injecting the genes later during development, a few days after fertilization. This may be done by introducing genes into individual cells that are then incorporated into the embryo before it implants in the womb.

The process of gene injection into an egg is quite tricky. The glass needle used to inject the gene solution is fragile and very, very fine; in fact, the tip is so fine it cannot be seen by the naked eye. Any untoward pressure on the needle breaks it.

Once the genes are safely injected into the nucleus, the egg, with its new complement of DNA, is transferred into the uterus of a foster female animal. The key to the potential power, and danger, of genetic manipulation is that we are changing the genetic code not only in the particular mouse that will be born of that one embryo, but changing the code for all its future offspring and descendants.

Of course, we are a long way from perfecting the technique. Firstly, the whole process is very labor-intensive, and the egg is often damaged or destroyed by the process of the injection itself. Frequently, the embryo does not develop properly after injection or, if it does, it miscarries during early development.

In many cases the embryo develops, but then there is no sign that the genes are present and functional, that is, the genes have not "caught," and the animal is not transgenic. Sometimes, the gene sequence becomes incorporated, the animal becomes transgenic, but the DNA does not express it, does not work normally to produce messenger RNA and the protein required. Sometimes gene expression starts temporarily, and then halts altogether. Not infrequently, expression occurs, but to a very limited and weak extent.

Transgenes can express in some parts of an animal and not others; some tissues might be completely unaffected by the new gene. If the gene in question is useful in, say, the liver, then the gene needs to be expressed in the liver. It is no good if the gene is only present in the nervous system. Sometimes, the rest of the animal's DNA is corrupted and the transgenic animals are born with severe congenital defects.

What is more, all these effects, or lack of them, are highly unpredictable. There would be no point in trying to cure cystic fibrosis in a human before birth, if, after a few weeks following breast-feeding, the disease started to rear its ugly head. If the baby also had other defects caused by the technology of gene injection, that would be worse still. We could never use it in humans until we can control the process completely and iron out all the creases. One scientist, Carol Readhead, had to inject 1,600 mouse embryos before the birth of a single mouse that properly expressed a particular gene she wished to study. Each transgenic pig, suitable for one human cardiac transplant, might well cost over $250,000.

There is one possible ingenious solution, and it promises to be a much easier and more reliable method. The idea is to transfer new genes into the animals' sperm instead of the fertilized egg. If this worked, it could revolutionize transgenic technology.

Sperm cells are the product of progenitor stem cells called spermatogonia. Throughout the life of a male animal, the testis filled with these stem cells is constantly producing new sperm, and at a huge rate. An average adult male, reading this chapter reasonably quickly, will have produced around 100,000 new sperm by the time he reaches the last period. The idea is that if we can persuade transgenes to enter these spermatogonia, then thereafter all the sperm they produce will each carry the code of the new transgene. In true Superhuman style, we would be using the body's own laboratory to spread the new gene among its progeny.

Currently, we are attempting this approach in various ways. One that holds some promise is to remove the spermatogonia from an animal's testes and then keep them alive in culture. Once in culture, there are various methods to transfer the new gene into the spermatogonia. One trick involves genetically modifying a virus to include the transgene in its DNA sequence. Then the spermatogonia are deliberately infected with this virus. Before the genetically modified sperm is transferred back, any sperm already present in the testes can be killed by radiation or drugs. This will prevent the modified sperm from being mixed up with the old unmodified sperm. After transfer, every sperm produced by the spermatogonia should have the new transgene encoded in its DNA. Whether we inject the transgenes into the fertilized cell, the embryo at a very early stage, or the spermatogonia, the transgenes find their way into all the cells and tissues, including the germ cells—the eggs and the sperm. Therefore, the genetic swap we make would be present not only in this first animal, but also in any offspring down the line. Theoretically, therefore, we might replace the mutated gene for beta-thalassemia or cystic fibrosis forever.

Testing the process on humans is a very long way off. But let's speculate. Fifty or a hundred years down the line we may be able to reduce the incidence of congenital disorders such as cystic fibrosis. We may also be able to reduce the genes that predispose people to certain kinds of cancer. There could be progress in detecting which genes predispose us to heart disease. A host of medical conditions and predispositions could be changed, forever.

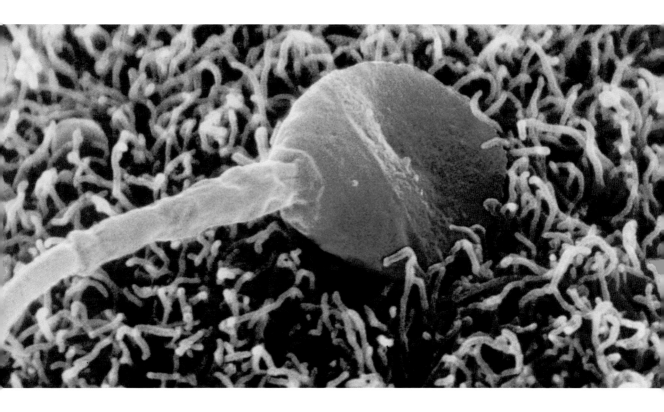

But it is wrong to suppose that these defects would be totally eliminated. That seems impossible because so many human genetic diseases occur as spontaneous mutations. For example, think about Duchenne muscular dystrophy. At least 30 percent of cases of this sex-linked inherited disorder arise as new mutations. A male child can show symptoms of the disease even though nobody in the family has been affected by it previously.

A human sperm about to penetrate the egg, magnified some 6,000 times. This process, when successful, takes a fraction of a second, but the fertilization that follows takes around 18 hours.

> *"Which brings us at last," continued Mr. Foster, "out of the realm of mere slavish imitation of nature into the much more interesting world of human invention."*
>
> *He rubbed his hands. For, of course, they didn't content themselves with merely hatching out embryos; any cow could do that. "We also predestine and condition. We decant our babies as socialized human beings, as Alphas or Epsilons."* [1]

1 *Brave New World*, Chapter 1, Aldous Huxley.

The flip side of this tempting technology, genetic modification, is its potential for creating "designer babies." It is just possible that there may be genes that predispose people to greater intelligence, beauty, or happiness. In that case, and if those genes could be pinpointed, there is no reason why genetic tweaks might not make a substantial difference to the mental and physical attributes of our children. This, of course, is at the heart of many people's fears about genetics.

Genetic manipulation of germ cells and sperm to prevent the ravages of serious disease is relatively easy to accept. For people who object to the embryo selection involved in PGD, there might be a moral advantage in treating an embryo at the beginning of pregnancy, rather than discarding it because it was abnormal. And even better, if we are attempting to improve health, why should we not treat sperm or egg cells before a child is conceived?

Most people seem to agree on the benefits of excising fatal congenital conditions from the germline. Why not go a little further still? We could give our children an even better leg up the genetic ladder. Why not engineer a slender physique and gleaming white teeth for our future models? Why not boost hand-eye coordination and lung capacity for our budding athletes? Why not inject a little high IQ for those future academics? After all, we give children orthodontic treatment and fluoride toothpaste to improve beauty; we take hormone replacement therapy to prolong youth; we ensure the maximum muscle strength of an athlete by training before the Olympic Games; we ensure that our soldiers in the army are fit for fighting and trained in aggression. Why not take the whole package and create a genetically engineered Superhuman with all those physical advantages permanently ingrained for generations to come?

There are many cogent objections. Firstly, genetic modification and the technology that surrounds it will inevitably be expensive, and only available to a moneyed élite in developed countries. This social and economic élite already has the benefit of a head start—private education and, more often than not, a comfortable financial cushion to fall back on. With genetic manipulation, they would also grab the genetic advantage. Their progeny would have better health, and, perhaps, be genetically designed for good looks and high intelligence. The presence of such an inherited genetic élite could only increase the divisions within society. It would surely reverse any progress we have made toward fairness and a society that promotes the development of all individuals within it. It would also be likely that the consumers of this technology would prize certain physical attributes above others—for example, strength, aggression, or intelligence—attributes that, wrongly prized, would cause great friction and upheaval for the generations to come.

Of course, even if we overcome the hurdles, the technical difficulties, of manipulating the germline safely, there is no guarantee that we could find the genetic keys to these incredibly complex physical and mental traits. No single gene that we have found determines IQ, height, or strength. Intelligence, beauty, and aggression may, indeed, have a powerful genetic basis but that probably derives from a complex web of genes that may be impossible to decipher. We also know that if we alter one gene on a chromosome, another gene some distance away from it may start to function differently, possible in a bizarre fashion, leaving a modified individual unexpectedly prone to disease or deformity.

Even a relatively simple condition, such as insulin-dependent diabetes, which is basically a lack of control of sugar metabolism caused by a lack of insulin secretion, could not easily be ironed out of the germline. We know of at least 20 different genes of several different chromosomes that are likely to predispose us to this disease. Unraveling the knots that constitute these genetic mechanisms may be impossible, at least for the foreseeable future, and even if we understand the mechanism, we may not be able to interfere without causing more problems than we solve.

Another serious issue is the fact that transgenic technology would be bound to produce mistakes, and the mistakes would be permanent. We would give a heritage to our children that was handed on because of our fallibility, a heritage which could destroy the basic comfort of their lives.

Artificial orchestration of our genetic code is a strange phenomenon in the context of evolution. The prospect of genetic manipulation on a large scale changes the baseline of human existence—our DNA, which has heretofore developed into its present form over millions of years of pushing and prodding by the forces of natural selection.

But let us suppose that eventually human ingenuity is able to get rid of the technical problems involved in genetic manipulation, the risk of unpredicted abnormalities and the malfunction of other genes that have not themselves been manipulated.

It is worth being fanciful. Let's suppose that that germline manipulation takes off in a big way. In 100 years' time we could be replacing 100 genes in any given embryo; over ten generations, with 100 genes changed in each generation, that would make a difference of 1,000 genes, artificially changed or added to the germline. Most scientists agree that we differ from chimpanzees by somewhere in the region of 5,000 genes. If we push "designer" babies to the limit, we are talking about creating a human that is as genetically different from the average human as we are from chimpanzees.

We have the potential to grasp the reins of evolution for ourselves. The concept of the Superhuman takes on an entirely new meaning. We are no longer talking about combining human biological talents and capabilities with our own technological wizardry in order to stay healthy and live longer. The genetic Superhuman is about reinventing "humans" as a new species.

Whether we are religious or not, the sanctity of human life is one of the core values of our morality. We all agree that this sanctity of life is imposed upon us for various reasons, not least because many of us believe in a kind of human divinity. We see ourselves made in God's image; this, above all, is why human life is sacrosanct. But if some of us cease to be human and become Superhuman, what then? Will ordinary humans become disposable?

> ... if some of us cease to be human and become Superhuman, what then? Will ordinary humans become disposable?

This notion of a Superhuman reverts back to an older meaning. Nietzsche's "Superman" was one who could reinvent himself and go beyond what the author saw as our human failings. He said, "Man is a thing which has to be surpassed." It was this idea that was coopted by those who championed eugenics. Ironically, the man who invented eugenics was Darwin's cousin, Francis Galton. He devised the name "eugenics" from the Greek word for "well produced" to describe the scientific study of racial improvement.

From the early 1870s, Galton expressed his desire to see the improvement of humankind's inherited characteristics. He hoped "to check the birthrate of the Unfit, instead of allowing them to come into being." This would be accomplished by sterilization of the so-called unfit. Equally, he wished to see "the improvement of the race by furthering the productivity of the Fit by early marriages and healthful rearing of their children." For Galton, the so-called science of eugenics was the mark of a civilized society.

These disturbing ideas are at the heart of much of our discomfort and fear of the genetic future. Even though no one is proposing sterilization of the "unfit," do we want to be turned into an idealized, possibly more homogenous humankind? I am not sure that we do. I like having flaws, and much of the time I like the fact that other people have flaws. It seems to me that our imperfections and differences are what make us individuals—and human. Of course, we want to live long, happy and healthy lives, but do we want to be perfect?

All advances in genetics pose impressive and threatening challenges both ethically and scientifically. But I do not agree with those who say we have opened a Pandora's Box. We have not had much time to get used to the possibilities of genetic manipulation and it is right that we should engage in a serious and far-reaching debate. However, I believe that we are capable of regulating, modifying, and controlling the new technology. We do not ban the car because of the greenhouse effect, nor the computer because of its potential for hackers to wreak havoc. Prometheus gave us fire, but, in the light of the Great Fire of London, or the destruction of Dresden in 1945, should we have refused that gift?

Throughout this book I have tried to shown how the science of medicine is becoming more intelligent. We also need to be intelligent in the way in which we apply these technologies, particularly when it comes to transgenes, cloning, and manipulating the germline.

Similarly, the technologies of self-repair, gene therapy, and new cures for cancer should not be bids for immortality. There might be individuals who want to live forever, but, as a society with a population that expects to live 30, 40, or 50 years beyond today's norm, we cannot. Some have predicted a world in which 80 percent of the population will be over 80 years old. Perhaps this prediction will come true, perhaps not. Medicine may be capable of achieving this arresting statistic, but we could not, at least for the foreseeable future, handle the consequences, and should not hold it as an ambition.

Superhuman medicine is dragging us, sometimes kicking and screaming, into a brave new world. But just like the great medical breakthroughs of the last century, gene therapy, tissue engineering, transplants, and fertility treatments should not really be designed to take us beyond the human, but to help us enjoy what it means to be alive, to live healthy, pain-free lives, and to help repair our bodies quickly and well when things go wrong—in other words, not to be Superhuman, but to be, simply, a fulfilled human.

Conclusion

"Those who see innovation as a bandwagon
should not forget its potential as a hearse"

The Superhuman is, of course, a fiction, but this fiction is an excellent metaphor for our hidden ability, a heritage handed down by evolution and a product of our genes—an occult nature that is so powerfully a part of ourselves that it has to be understood if we are to have the best protection when threatened by disease.

In all aspects of human healing, the Superhuman is there, waiting for employment. After trauma, the genes that regulate injury, flight, and fight are crucial to our immediate survival, and the process of self-repair is an example of how the genes that control our development might be manipulated to grow new tissue and organs. Transplantation of healthy organs to replace those that are diseased will be immeasurably facilitated by understanding the genetic instructions to our immune system and understanding this genetic control of immunity will also be crucial in keeping the upper hand against bacteria and viruses. The fight against cancer, for example, can move forward only with an understanding of the Superhuman.

Currently there is one medical field where the Superhuman seems less important—reproduction. The scientific basis for our treatment of fertility disorders is poorly developed. In vitro fertilization is still a blunderbuss therapy and alternative treatments are seldom offered. The impetus to improve diagnosis and treatment seems poor; but more importantly, we know little about the genes that influence fertility.

Paradoxically, reproductive medicine is the field that holds the greatest genetic threat. Manipulation of reproduction could change the human into someone, who is truly superhuman. The unanswerable question this raises is this: If, in the future, we are no longer made *only* from the complement of human genes that defines our species, if some of us are enhanced genetically, will we still be human?

Much of the debate swirling around the Superhuman is, however, more mundane. Whether we talk about transplants, cancer research, or the search for AIDS vaccines, the basic issues, which are largely social and political, are mostly the same. How far are we willing to go to improve medicine? What price are we willing to pay, both financially and morally? How do we avoid unequal delivery of what seems an inalienable right—the chance of good health?

Some of the arguments are about risk. Should we risk the lives of innocent bystanders by using a seemingly harmless virus during gene therapy? Could a genetic database have implications for the privacy of ordinary people?

Much of the debate concerns issues related to public education and understanding. For example, does an animal's organ transplanted into a human body violate some basic law of nature? Is a person with a baboon heart still human? In offering more people a chance of life are we contravening some basic law of nature?

Some of the debate involves the very basis of our humanity. Is it right for humans to breed animals for research or to harvest their organs for transplantation? Are fertilized eggs sacred and should we do experiments with them to improve human fertility treatment? Should the cloning of embryonic tissues be allowed so that we can create new sources of tissues and organs?

Some physicians and scientists are affronted that such questions are even asked. There is an assumption that they know best, that they do not need to answer what seem to them to be ignorant questions or questions born out of fundamental misunderstanding. Fortunately, this approach is on the wane. The impact of nuclear fusion left a question mark over the field of physics; the threat of global warming, a cloud over advances in chemistry; thalidomide, an anxiety about the drug industry; and cloning and genetically modified foods, a pall over biotechnology.

Some people believe that science leads us into areas that are inherently dangerous. Certain areas of science, they argue, have hidden moral dangers and should not be explored. I believe that science itself is morally neutral. It is taking from the Tree of Knowledge, a God-given human gift inherent in human inquisitiveness, intelligence, and invention. Science itself has no moral value. It is the use of science—its exploitation—and the way that scientific knowledge is obtained that give rise to moral problems and dangers.

Having said that, it surely would be wrong to limit science to what is perceived to be free of moral risk, nor should we limit scientific exploration to areas that seem free of danger. If society decided on that course, inevitably some immeasurable, and possibly irreplaceable, discoveries of lifesaving benefit would be lost. Penicillin was not predicted from the mold found by chance on the windowsill of Nobel Prize winner Alexander Fleming. Basic science cannot forecast what its exploration will bring and chance finds lead to important benefits as well as potentially harmful risks.

Some people in our society would ban much of the high technological development that contributes to human health. The Human Genome Project, genetic screening, the manipulation of animal and human genes, and, above all, genetic engineering, are all perceived to be dangerous areas. However, banning any technology or indeed any human activity will not stop experimentation. Banned activity tends to continue, to become clandestine and consequently more dangerous. Had the Manhattan Project, the secret race to make the first offensive nuclear device, been public, and had the inner circles of American and British government been answerable to external authorities—the Cabinet of the day, Parliament, Congress, and the electorate—it is doubtful that we would have produced the most destructive weapon of all time. The bomb was spawned in secret.

On the other hand, a free-for-all is not a viable option for a healthy society. Regulation of technological processes is needed. Those of us living in the developed world are in a unique position to take this lead, a lead that must be seen to be instigated not purely from self-interest. The regulation needs to be preceded by proper public consultation, as it was, for example, by the Warnock Committee, which reported on the status of the human embryo in Britain before the government regulated in vitro fertilization by establishing the Human Fertilization and Embryology Authority.

This regulation process has been well established in the UK – for instance, there was the 1997 publication of the Kennedy Report on the Ethics of Xenotransplantation, which led to the UK Xenotransplant Interim Regulatory Authority and opened the door to serious research into the use of pig organs in transplantation. The Human Genetics Commission, which examines genetic issues in Britain, is another example of a body that is likely to set the scene for legislation and regulation in that field. In the United States, bodies such as the National Bioethics Advisory Commission should pave the way to a better public policy.

A major problem, however, remains. The most technologically advanced society of all, that of the United States, finds difficulty in regulating technology. There is a mistrust of central government, of orders from Washington, DC, which has existed ever since the Constitution was enacted. While federal bodies, such as the Food and Drug Administration, or FDA (page 113), rigidly ensure protection where pharmaceuticals and medical devices are concerned, individual states also have considerable autonomy.

Likewise, many aspects of scientific and medical research are not so well regulated. For example, work on a human embryo before transfer to the uterus is not illegal. Local ethical approval is required but provided the work is not funded by central government, there is no federal regulation. Similar procedures are not countenanced in Britain and are regulated by law. In the United States, however, there are many experimental procedures involving human subjects for which federal funding is illegal, but which, if the financial support comes from other sources, are permitted. In a sense, the federal government "washes its hands" of responsibility for what happens.

The American system works well when there are no serious accidents. It allows the pioneering spirit and entrepreneurial activity for which the United States is rightly famous. However, biotechnology is under the greatest scrutiny throughout the world at present. It affects the health and well-being of individual humans, and adverse threatening effects are immediately intelligible. In the UK, we have already seen how easily a valuable biotechnology, such as genetic modification of foods, can be brought to a complete stop if the public perception is that a process is dangerous even if the danger has not been proved. The furor over genetically modified foods is a prime example of this.

There is another problem—financial interest. Biotechnology is an area of science that is rapidly being commercially exploited, and this pursuit of vested interest deeply worries people. In the United States, commercial motives are often given a priority that seems unacceptable elsewhere. American exploitation of intellectual property is also increasing and the power of big American finance causes great concern both at home and abroad. Some doctors and scientists are seduced into taking part in this drive for profit and, consequently, it is increasingly difficult to separate their desire to help a sick patient from the possible advantages of being the first and most famous person in a high-profile field of medicine.

A good example of this conflict is seen in the recent events surrounding gene therapy. Gene therapy is portrayed as the Promised Land of Medicine, the ultimate in helping the Superhuman, involving as it does the transfer of genes that may successfully regulate disease processes. But it is high risk. As well as curing cancer, it could cause it. Gene transfer most often uses a virus as a vehicle and the use of viruses brings a risk of infection. New genes inserted into our DNA are unpredictable in their effect, not least because they may alter the function of other genes already present. Given that it promises potential cures for many fatal diseases, including cancer, gene therapy is naturally high-profile and it is hardly surprising, therefore, that there is great pressure to develop and exploit it commercially.

Some 5,000 patients have now undergone different forms of gene therapy for a variety of very serious diseases, all of them genuinely life-threatening and mostly after all other possible treatments have failed. But the recent death of teenager Jesse Gelsinger is a kind of watershed. When this 18-year-old died in the year 2000—his death was the first mortality attributed to gene therapy itself rather than to the underlying disease process—the field was halted. Moreover, many people doubted that his gene therapy was really needed. Jesse had a rare genetic disorder called ornithine transcarbamylase deficiency, a genetic liver disease that causes a buildup of ammonia in the body.

In most people, usually infants, this disease causes sickness, coma, and death, but in some individuals, it is controlled by a low protein diet and drugs. In Jesse Gelsinger's case, he was an active, athletic young man in no danger from his disease provided he ate prudently and took his pills. It is difficult to see how, in this instance, gene therapy could be justified but he volunteered to be a guinea pig; and, more to the point, was accepted as one by the medical practitioners responsible for his welfare.

Jesse's treatment was partly under the supervision of Dr. Jim Wilson, of the University of Pennsylvania. By many accounts, Jim Wilson is a highly focused, ambitious physician who has been chasing his elusive goal of successful gene therapy. Wilson and his colleagues had done many experiments on mice and monkeys and they persuaded the powers that be that the time was right for testing this treatment on people.

Jesse Gelsinger agreed to gene treatment, it is reported, because he wanted there to be more knowledge to help very ill babies with his disorder. His death occurred after an injection of a larger-than-average dose (in a randomized trial) of the adenovirus carrying the desired corrective gene. His death, protracted over four days, was horrific. According to newspaper reports, he rapidly developed a high fever (over 104°F), became jaundiced, the blood in his vessels slowly coagulated, his lungs gradually failed, he swelled up all over, and, little by little, the functioning of all his vital organs ground to a halt.

This seems to be largely what happened to three of the trial monkeys given similar high doses. But it seems that neither Jesse Gelsinger nor his parents were told of the monkeys' deaths before the signing of the consent form. Many people find it difficult to escape the view that in Jesse's case the researchers may have been prepared to take life-or-death risks—for glory. Of course, it was the patient's life or death for the researchers' glory.

At the time of writing, the trials at the University of Pennsylvania where Jesse Gelsinger was treated, are suspended. The National Institutes of Health (NIH) is conducting an investigation, and, no doubt, questions will continue to be asked at the highest level in Washington. They certainly are in Britain.

"As a result of Jesse's death," stated his father, Paul L. Gelsinger, during his testimony given to the US Senate's Health, Education, Labor and Pensions Committee Subcommittee on Public Health, "many important issues regarding gene therapy have come to light. The number and lack of proper reporting of adverse events associated with gene therapy, the secretive nature of gene therapy research, and the motivations behind the race for results are what trouble me most."

Jesse Gelsinger – was he a victim of medical over-enthusiasm?

The National Institutes of Health in Washington quickly called for a full accounting of all the adverse events in gene therapy trials, whether or not these had been found to be related to the therapy itself. According to one report, there have been 691 incidents involving death or illness after various kinds of gene therapy, but only 652 have been reported promptly as required by federal rules.

Among those incidents, which suddenly came to light, was the death of Roger Darke. Cardiologist Jeffrey Isner of St. Elizabeth's Medical

Center in Boston, in whose VEGF gene therapy trial Darke died—the same trial from which Charles Wilson (see page 107) emerged pain-free and healthy—took some of the blame. But Darke's death had been determined at autopsy to have been caused by his disease, not by the treatment that carried the genes to his blood vessels. The death, although not reported to the NIH, had been reported to the US Food and Drug Administration, as required.

During his Senate testimony mentioned above, Paul Gelsinger said, "I read that my son's death was called by one of the leaders in this field 'a pothole' in the race to gene therapy. But his death was no pothole. It was an avoidable tragedy from which I will never fully recover. My concern now is that Jesse's death will not have been in vain, not be just a pothole. I am not against gene therapy. I am all for its continued development, but it must be better regulated. I recognize it holds so much promise for so many people. But we cannot allow what happened to Jesse to happen again."

Leaving aside the appalling aspects of Jesse's death, this abortive attempt at pushing the frontiers too far too fast had other serious effects. Gene therapy that could have been of value to patients who will otherwise inevitably die has been halted in the United States. That, for example, includes the mesothelioma trial that cancer specialist Daniel Sterman, also from the University of Pennsylvania, was conducting, the one that Bryan Newton (see page 160) had entrusted with so much of his hope, because there was no other treatment. Moreover, whatever happens scientifically in the United States inevitably spills over into Europe, particularly the UK. Gene therapy worldwide has suffered a serious blow, and with it the hopes of many potential patients. There are important lessons to be learned from the tragic story of Jesse Gelsinger.

A recent report of the Select Committee for Science and Technology of the House of Lords pointed out that "Society's relationship with science is in a critical phase. Science today is exciting and full of opportunities. Yet public confidence in scientific advice to Government has been rocked by BSE; and many people are uneasy about the rapid advance of areas such as biotechnology and IT, even though for everyday purposes they take science for granted. This crisis of confidence is of great importance to British society and to British science."

For BSE, one could just as well read gene therapy or indeed any advanced medical development involving human genetics. Given the American mishaps, for British science one could easily read American science.

It is right that there should be real concern about the relationship between science, scientists, and the public. Without more recognition of what is a growing problem, both science and medicine could experience a serious public backlash, one that could have a profound effect on our well-being and our viability as a productive modern nation.

Scientists do not always engage the public and do not always seem open about what they are doing. During the making of the television series that accompanies this book, it was notable how often British physicians and scientists were reluctant to get involved with film cameras. By comparison, their American counterparts were all too ready to show their wares, except when they were likely to face really serious public criticism—for example, over the issue of hand transplantation.

We need a happy medium. Undoubtedly, scientists should do more to contribute to public education. Open communication with the media, which is very much a feature of science in the United States, is much less so in the UK. The American model has much to be commended. However, I do have some reservations about the American approach, which too often seems to be born out of a sense of self-aggrandizement, or to advertise something which is of advantage to the scientists. This seems more frequent where medicine is practiced privately.

We also need to find better ways of conveying the excitement of science to schoolchildren, especially the very young. In Britain, science teaching in primary schools is all too often delegated to arts graduates as a sop to the curriculum. Areas of the science curriculum need to be put into a more social context and given more relevance to the problems of society. Children learn about the periodic table—the classification of the elements—but the relevance of this to radioactive waste is seldom taught. They learn about DNA, but have little understanding of its relevance to human medicine.

As mentioned before, vested interests in biotechnology and medicine remains one of the biggest concerns. And vested interests, it has to be said, are widespread and growing. When I was a young research worker, the thought of patenting the medical devices we had made was totally alien. We felt they should be available for the public good. Now it is mandatory to ensure the maximum financial rewards. The growth of intellectual property and the patent industry, the hunger of the universities, the ambition of their employees and their need to find supporting funds, the growing expense of medical care—each contribute to this.

All these considerations are relevant to the case of Jesse Gelsinger. It is easy to see how further gene therapy disasters could cause a radical change in public opinion, and it must always be remembered that it is public opinion that drives progress in medicine and science. The public pays the bill. Scientists need to be less arrogant; they need to recognize that they are the servants of society, not its masters.

At the beginning of this book, I pointed out that the principle of promoting the best conditions for the body to heal itself—allowing the Superhuman to work—is still relevant long after Hippocrates' death. I took pains to explain that this was not a plea for so-called alternative or complementary medicine, but I do believe that doctors would do well to practice medicine with a holistic approach. The great breakthrough promised eventually by knowledge of the human genome means that we shall be able to tailor treatments—and drugs as well—to the genetics of the individual. However, paradoxically, we shall also need to be more modest about these genetic achievements of ours. The incredible hype that has surrounded the recent publication of the first draft of the sequence of the human genome is a case in point. The fact that we now have a printout of that unintelligible tangle of as yet untranslated code is actually, in itself, of no value to anybody. For "responsible" scientists to claim that it is like the invention of the wheel is to stretch public tolerance of announcements of scientific "breakthroughs" to the limit.

A truly holistic approach must involve more than merely gaining the confidence of individual patients. More than ever before, medical treatments, with science behind them, require the confidence of society. Practitioners will need to be more responsive and less authoritarian. They will need to be open and to demonstrate good faith because modern medicine raises too many human issues, which affect both the healthy and the ill, for any other approach.

Carved in the lintel over the entrance to the medical school where I trained is written a perfect example of the holistic principle. It is a motto from Terence, the Roman author and playwright, who died around 159 BC.

"*Homo sum. Humani nil a me alienum puto.*" A rough translation is: "I am a man. I consider nothing human is alien to me." In the practice of medicine, while we may wish to consider the Superhuman in our patients, we must always remember that we doctors are merely human.

Index

Picture credits

BBC Worldwide would like to thank the following for providing photographs and for permission to reproduce copyright material. While every effort has been made to trace and acknowledge all copyright holders, we would like to apologise for any errors or omissions.

BBC graphic images by Gillian Best, Ina Hurst, Rita Kunzler and Rick Leary.

10 Just Greece Photo Library; 12a The Bridgeman Art Library; 12b Science Photo Library; 15, 17, 24–5, and 29 Science Photo Library; 30 Oxford Scientific Films; 32 Science Photo Library; 33 The Art Archive; 34 The Wellcome Trust; 35 Hulton Getty Picture Collection; 37 National Motor Museum; 39 www.shoutpictures.com; 41 Science Photo Library; 45 Robert Winston; 47 PA News; 56 Oxford Scientific Films; 57 Science Photo Library; 62-3 BBC; 64 Robert Winston; 67a Peter Liuzzo; 67b Getty/Stone; 69 John Groom; 71 The Bridgeman Art Library; 77 Science Photo Library; 83 John Groom; 84a Sygma; 84b Sygma; 87a Wellcome Trust; 87b Wellcome Trust; 95 Associated Press; 97 PA Photos; 98–9 BBC; 101 Steve Nolan; 103 BBC; 110 and 113 Science Photo Library; 117 Associated Press; 120 Robert Winston; 121 and 125 Science Photo Library; 126 Dr Carol Readhead; 132–3 BBC; 135, 136, 138, 139, 139b and 143 Science Photo Library; 146a Getty/Stone; 146b Science Photo Library; 147 Science Photo Library; 148 Science Photo Library; 151a Science Photo Library; 151b Getty/Stone; 153 Science Photo Library; 155 Mike Coles; 159 and 163 Science Photo Library; 166 Robert Winston; 168–9 Science Photo Library; 171 The Bridgeman Art Library; 172 and 173 Science Photo Library; 174 Tim Streatfield; 175 and 177 Science Photo Library; 181 and 182 Liesel Evans; 183 Science Photo Library; 184 Robert Winston; 189 Robert Winston; 191 Robert Winston; 193 Science Photo Library; 198 The Imagebank; 200 Robert Swan; 202–3 BBC; 205 The Bridgeman Art Library; 206, 207 209 and 211 Science Photo Library; 213 The Imagebank; 217, 226, 227 and 228 Science Photo Library; 234 Robert Winston; 236 and 239 Science Photo Library; 249 Gelsinger family.